José Chiumbo Paiva

Foundations and requirements of teacher preparation

José Chiumbo Paiva

Foundations and requirements of teacher preparation

for the development of Mathematics Olympiads

ScienciaScripts

Imprint

Cover image: www.ingimage.com

This book is a translation from the original published under ISBN 978-620-2-15416-1.

Publisher:
Sciencia Scripts
is a trademark of
Dodo Books Indian Ocean Ltd. and OmniScriptum S.R.L publishing group

120 High Road, East Finchley, London, N2 9ED, United Kingdom
Str. Armeneasca 28/1, office 1, Chisinau MD-2012, Republic of Moldova, Europe
Printed at: see last page
ISBN: 978-620-7-75665-0

Contents

About the author

Jose Chiumbo Paiva was born in the province of Huambo in the Republic of Angola. He graduated in 2013 in Educational Sciences, specialising in Mathematics, at the Higher Institute of Educational Sciences of Huambo (ISCED-Huambo). In 2017 he obtained a Master's degree in Applied Mathematics at the Central University "Marta Abreu" of Las Villas (UCLV) in Cuba. Since 2017 he is a doctoral student in Educational Sciences at the UCLV.

She joined the Education Sector in 2011, and currently teaches at the Escuela de Magisterio "Ferraz Bomboco" in Huambo, where she regularly teaches Mathematics, Mathematics Teaching Methodology and Teaching Practice.

His teaching experience includes a stint at ISCED-Huambo as a collaborator, where he has taught Functional Analysis, Differential Geometry, Practical Problem Solving in Elementary Mathematics, Mathematics I and II.

DEDICATION

To my parents;
To Evalina Chilombo, my wife;
To my children;

To my brothers and sisters

Acknowledgements

I wish to thank MSc. Alambre Jose Pinto for his cooperation in the preparation of the teachers for the development of the Mathematics Olympiads, especially Dr. Tomas Pascual Crespo Borges, Dr. Eric Tomas Crespo Hurtado and Dr. Guillermo Soler Rodriguez, for their very valuable recommendations and criticisms.

Summary

At the international level, Angola's results in the Mathematics Olympiad are very poor, and at the national level, in the province of Huambo, these competitions are not carried out with the required systematicity. This is largely due to gaps in the theoretical and methodological training of teachers.

This book provides the fundamentals and requirements for the preparation of teachers in order to improve their performance in the development of the mathematical olympiads.

Initial comment

Education has historically assumed the mission of preparing man for life, according to the social task set. At present, the world is going through a deep economic crisis, a situation which demands the training of a man who can quickly and creatively find solutions to the political, social, economic, scientific and environmental problems of these times.

It is the teacher who is responsible for contributing to the ideological development of his or her students, to ensure that they play a leading role in all school and extracurricular activities, so that they become people capable of keeping pace with the new times.

The educator has to be prepared to meet new personal and social needs, and to know how to face and promote initiatives in the face of new contradictions. The preparation of teachers, and consequently the improvement of their professional performance, is a highly topical problem of international relevance, as the literature review has shown.

In response to this need, the Angolan government increased the number of teacher training colleges and the capacity to provide pre-service training for the new generation of teachers.

Despite all the progress and efforts made by Angolan education and the professional preparation of its teachers, the proposed objectives, in which professional preparation must have a special place, have not been achieved. Teachers are faced with the challenge of finding ways to guarantee the learning of mathematics in schoolchildren, in order to face the challenges and solve the problems of the different social spheres in a globalised world with a strong crisis and high development in ICT.

Knowledge and skills competitions have proven to be an effective means of promoting interest in learning and raising the quality of the teaching-learning process in all types and levels of education.

For this reason, the first National Mathematics Olympiad was held in the country in 2010 with the intention of achieving a progressive increase in the quality of education. However, there are gaps in the theoretical and methodological order for the Angolan context in this important training process of teachers in order to improve their performance for the development of the competitions.

This is a requirement for Angola's current aspirations in the education sector, so that despite the efforts made and the changes that have been introduced to achieve better results in the didactic performance of teachers, the requirements for the preparation of teachers and the new generations are still not being met through the role of competitive examinations in the educational field.

Although a regulation for the development of the "Olimp^adas de Matematica" was approved and updated in 2018 in the Executive Decree No. 03/18 of May 15 of the Ministry of Education, the compliance with the legislation is left to the spontaneity of the teachers, from the development of the competitions to the preparation of the participants.

Article 4 of the Regulations for the "Mathematics Olympiads" expresses the objectives of this competition, which recognises the importance of the teaching of Mathematics, the role of the competitions in motivating students to study this discipline, in detecting young talents, in improving the quality of teaching and learning of Mathematics, among others.

However, the teachers in the educational establishments responsible for this task do not give it due attention. This is due to a large extent to the lack of guidelines and guidelines that would serve as a basis for achieving the objectives of the "Mathematics Olympiads".

This is also influenced by the fact that the teachers who teach the discipline of Mathematics and who are in charge of running the competitions have not had any pedagogical or didactic training in Mathematics.

This is a latent problem that conspires against the very development of the competitions from the selection and preparation of talented pupils. The following aspects constitute the main difficulties:

S Quiz tests are the main instrument for identifying talented pupils; moreover, their structure and rigour of problems is the same regardless of the level of competition.

S The exercises proposed both in the tests and in the preparation of the contestants stimulate more calculation than problem solving, and intra-mathematical curricular content prevails over problems that stimulate creativity.

S There is no classification of the problems that would allow the teacher to make a selection of these for the preparation of the tests and in the preparation itself.

S Teachers do not follow up on students who have won at one of the stages of the competition and do not use adequate forms of preparation.

S In the various stages of the first phase of the "Olimpiada de Matematica" no reports are produced and in the other phases reports are produced with administrative characteristics without mentioning the main difficulties of the contestants in solving the exercises.

The above reveal the existence of contradictions between the aspirations of the Angolan Ministry of Education, and the results that are being achieved in the activity of Mathematics competitions motivated mainly by the lack of preparation of teachers for the development of these. Consequently, with the above, the author has outlined in this book the fundamentals and requirements of the preparation of teachers for the development of Mathematics Olympiads.

The book consists of five ep^graphs, where the theoretical-methodological conceptions on tho teacher's preparation in general and its particularities for the development of Mathematics competitions are exposed, as well as the background of these competitions in Angola. A teaching experience of the author concerning the preparation of teachers for the development of Mathematics competitions is presented and the main components of the requirements of the preparation of teachers for the development of Mathematics Olympiads are addressed. The final part presents the challenges and changes needed to improve the performance in mathematics competitions in the context of the Huambo province of the Republic of Angola.

CHAPTER 1

Mathematics competitions and Olympiads, background and rationale

The knowledge olympiads are healthy and educational competitions that take place in almost all countries, where the academic preparation and effort of the students is evaluated. Among them are the Mathematics Olympiads, which, having an international scope, the participating countries can be grouped by region, such as the Ibero-American Mathematics Olympiads, or the Mathematics Olympiad of the Community of Portuguese-speaking Countries (OMCPLP), which brings them together by language.

So the Mathematics Olympiads can be considered as a competition equivalent to a sports competition, where the "athletes" are students and the "coaches and trainers" are the teachers. For any competition the "athletes" receive a specific preparation, which consists in this case, in solving mathematical problems individually or in teams, and for this the "coaches and trainers" dose the training so that their athletes develop logical skills of mathematical thinking, creativity and sociability for teamwork.

In order to participate in international Olympiads, young people and adolescents have to go through the competitive cycle in their country of origin, which develops in a pyramidal way until it culminates in the competition at national level. In this sense, Hungary is considered the initiator of these competitions in 1894 and since then until today they are held in more than a hundred countries, organising mathematical olympiads, or similar competitions with different names.

All these competitions, to a greater or lesser extent, have as their primary objective to stimulate the study of mathematics and the development of young talents in this science and in school practice. The knowledge competitions constitute a motivating component for pupils, teachers and professors to study and study in depth the subjects in which this activity is organised, hence some of the following expected objectives are recognised:

- To encourage the massive participation of pupils in the activities of the competition, mainly in the classroom and school stages.
- To develop an interest in study by systematising, broadening, deepening and consolidating the knowledge and developing the skills set out in the programmes.
- Provide incentives that are transformed into pleasurable mental activities and stimulation of the intellect.
- To promote the quality of learning and the improvement of the promotion rates of the subjects.
- To morally stimulate the effort and work of teachers and students.

Today the Mathematical Olympiads are widely known, both in the mathematical community and in the community at large, for the impact they have had and continue to have in transforming the way students perceive themselves through their growing confidence in mathematical knowledge, creative power and the strength of their thinking. No less important is how we, teachers and professors, perceive our actions as trainers, to what extent we have been able to transform our students in their preparation, both in mathematical knowledge and in their personal formation to integrate into society.

The Olympiads have come to take different forms, from quick multiple-choice tests to investigative tests of several weeks' duration, composed of tasks that adjoin or lead to open-ended problems. Regardless of the form and size of the problems, mathematics is sufficiently broad and flexible that all of these formats allow for problems that stimulate the student's ability to develop self-improvement in mathematics. Even multiple-choice tests (which make it possible to organise mass participation competitions) give every student the opportunity to solve simple but intriguing problems set in familiar circumstances.

The ways in which the Mathematics Olympiads and their different modalities have been developed have evolved throughout its history, but always with the aim of stimulating the love for this science.

CHAPTER 2

Historical background of the development of Mathematics competitions in Angola

Angola achieved independence in 1975 with an illiteracy rate in the order of 85%, one of the highest in the world (UNDP-Angola, 2002). This was the result of the segregationist educational policy of the colonial regime that had been in place for almost five centuries. This dramatic inherited situation led the new government to establish the Education Sector as one of the priorities of its policy, as a means to consolidate national independence and defined it as a right based on the principles of universality, free access and equal opportunities.

This allowed the reduction of the illiteracy rate in the country, as the final results of the 2014 Census (the only one carried out after national independence) show that Angola has a literacy rate of 66%, with the urban area with 79% and the rural area with 41%. Likewise, another official government document published in April 2018, called the National Development Plan 2018 - 2022, states that in that year the literacy rate was 75%.

As the education sector was one of the priorities of the new government's policy, it generated a school explosion. According to (Liberato, 2014), while prioritising education was an achievement, mass access made educational management more complex due to the accelerated degradation of school infrastructure. Numerous schools were closed, as were libraries, sanitary facilities, gymnasiums and cafeterias. The situation was aggravated by the civil war, which had a negative impact on all sectors of the country, leading to the questioning of the quality of the education system, especially considering that most teachers were not trained for the job.

In the National Development Plan 2018 - 2022, which has already been referenced, the Angolan government recognises that "the secondary education sub-system (grades 7-12) faces several challenges, such as insufficient classrooms and teachers to meet the demand for education, as well as poor infrastructure. The lack of textbooks and teaching materials also weakens the quality of the education provided. (UNDP-Angola, 2002)

However, between quality that justified the selective and discriminatory nature of education in the colonial regime and quantity that guarantees equal opportunities for all citizens to access schooling, the population opts for the latter. They prefer to have pupils take lessons outdoors in the shade of a tree, sitting on the ground or sheltered in an abandoned porch, rather than having a select few pupils in conventional, technologically well-equipped classrooms (Mazula, 1995) cited by (Liberato, 2014).

The entry into the new millennium brought with it new policies for the education sector in Angola. In September 2000, after the so-called Millennium Summit, which brought together 189 countries that signed this document, Angola started a "profound process of revision of the policies and strategies that regulated the sector" (UNDP-Angola, 2002), which led to the elaboration of the Integrated Strategy for the Improvement of the Education System (2001-2015) and the approval in 2001 of the Basic Law of the Education System, Law no. 13/01, of 31 December, the latter being updated in 2016 by Law no. 17/16 of 31 October, which was approved by the Ministry of Education and Culture.13/01 of 31 December 2001, the latter being updated in

2016 by Law n°.17/16 of 7 October, and amended again in 2020 by Law 32/20 of 12 August. These two documents set out the reforms to be implemented in the education sector, the general objectives of which are: 1° Expand the school network;
2°. Improve the quality of education;
3°. Strengthen the effectiveness of the education system; and
4°. Establish equity in the education system.

Regarding the improvement of the quality of education, the reference documents state:

- In-depth reformulation of the general objectives of education, school curricula, contents, pedagogical methods, structures and appropriate pedagogical means;
- Improvement of the pedagogical and learning environment for students;
- Initial and in-service teacher training;
- Modernisation and strengthening of school inspection;
- Improving the quality and quantity of textbooks;
- Improvement of methodological work and teaching;
- Ensuring community participation in school work, i.e. guaranteeing the relationship between the school and the community;
- Reduction of illiteracy;
- Expansion of the catch-up programme.

They also structure the Education and Education System, which is unified and consists of four levels of education and six sub-systems of education.

The levels of education are:

- Pre-school education;
- Primary Education;
- Secondary Education;
- Higher Education.

The Education Subsystems are:

- Pre-school Subsystem;
- General Education Subsystem;
- Subsystem of secondary technical-vocational education;
- Teacher Training Subsystem;
- Adult Education Subsystem;
- Higher Education Subsystem.

The General Education Subsystem is the basis of the Education and Education System. It is structured into Primary and Secondary Education.

General Secondary Education is the level that follows Primary Education and aims to guarantee a comprehensive, harmonious and solid education, necessary for a good insertion in the labour market and in society, as well as for access to subsequent levels.

General Secondary Education comprises two cycles of three classes and is organised as follows:

S The I Cycle of General Secondary Education comprises grades 7, 8 and 9 and is attended by students who are at least 12 years old in the year of enrolment;

S The II Cycle of General Secondary Education comprises grades 10, 11 and 12 and is attended by students who are at least 15 years old in the year of enrolment.

As mentioned above, one of the social objectives addressed by the government in Angola is the progressive increase in the quality of education. It is for this reason that in 2010 the first national mathematics competition was held in the country, called "Mathematics Olympics".

At the beginning of July 2011 began in Coimbra - Portugal - the first Mathematics Olympiad of the Community of Portuguese-speaking Countries (OMCPLP), where Angola is a regular participant. The objectives of these competitions are:

- Improving the quality of teaching and discovering talents in mathematics, especially for scientific and technological development.
- To encourage the study of mathematics in the "Lusophone" countries.
- The creation of opportunities to exchange educational experiences between nations;
- The union and co-operation between the "lusophone" countries for the creation of instruments to allow the competition of pupils in an international olympiad for Portuguese-speaking countries.

Although the first Mathematics Olympiad in Angola was held in 2010, the first national pre-selection took place in 2011 in response to the need to send Angolan students to participate in the OMCPLP starting that year.

From that moment on, a process of improvement of the Angolan Mathematics competitions began. It gains in quality every year in terms of organisation. For the development of these competitions, a regulation was approved in 2015 in the Executive Decree n°142/15 of 26 March, and updated in 2018 in the Executive Decree n° 03/18 of 15 May, both of the Ministry of Education.

The objectives of the national competition "Olimpiadas de Matematica" are set out in Article 4 of these regulations and are as follows:

a) Recognise the importance of teaching mathematics;
b) To motivate students to study the discipline of Mathematics.
c) To contribute to the improvement of mathematics teaching.
d) Detecting young talents in mathematics.
e) To create opportunities for exchange of experience in the field of mathematics.
f) Selecting students to participate in international mathematics competitions
g) To improve the quality of teaching and learning of mathematics, primarily for scientific and technological development.

The national competition "Olimpiadas de Matematica" is aimed at pupils of public and public schools of primary education and the first and second cycle of secondary education. This means that the age of the contestants ranges from 11 to 16 years, corresponding to the 6th grade (pupils aged 11), 9th grade (pupils aged 14) and 11th grade (pupils aged 16).

According to these regulations, the organisational regulations for this activity are established from the classroom to the national level, where the Ministry of Education is responsible for the development of the activity. Each province can apply to host the final phase, which must include the provincial directors and the winners of each category (6th, 9th and 11th grade). The provinces will be represented by six students accompanied by a teacher (Mathematics Discipline Coordinator).

The "Olympiads of Mathematics" are organised in a staggered manner in three phases:

- **1st Phase:** Corresponds to the competitions at classroom and school level, between different schools in a municipality and between the winners of each municipality (provincial stage).
- **2nd Phase:** Corresponds to the national pre-olympics where the winners of each province compete to select 18 students to compete in the next phase.
- **3rd Phase:** 21 contestants participate, being 18 students who passed the pre-olympics plus 6 from the province hosting the event. In this phase a total of 9 winners are selected, 3 for 6th grade, 3 for 9th grade and 3 for 11th grade. From among the top performers of the 11th grade group in this phase, students are selected to participate in international Mathematics competitions, especially in the OMCPLP which Angola has been attending regularly since its inception in 2011.

For each of these phases, the responsibilities and functions of the different directors, the Provincial Director or Secretary of Education, the Municipal Director of Education and the Director of the Public or Private School are defined. These responsibilities are at an organisational level and include the promotion and dissemination of the Mathematics Olympiad in accordance with the level, forming the tribunals, guaranteeing the transport of the municipal or school teams, carrying out the opening ceremony, delivering the results of the winners to the higher level and disseminating them according to the established deadlines, among others.

In the particular case of the teachers of Primary Education and those of Mathematics corresponding to the I and II Cycle of Secondary Education, they are responsible for them:

a) Explain the regulations;

b) Motivate students to participate in the competition;

c) Instruct students on the rules to be observed during the preparation and conduct of the competition.

As it can be seen, the functions and precisions given to the teacher for his performance in the development and preparation of the students for the Mathematics Olympiad are very brief.

With regard to the composition of the tribunals at school level and their functions, it is proposed that they are made up of three teachers, one of whom is the president and the other two are members. The panel is in charge of drawing up, applying and marking the test. For the test, they must draw up a proposal of 6 problems for the second phase, which must be sent to the municipal board for approval. They are also responsible for disseminating the results and sending a report on the 1st phase.

The Municipal Commission or Tribunal for the competitions is chaired by the Head of the Education Area and by the Municipal Coordinator of the Mathematics discipline. A person in charge of the Out-of-School Activity of the Country Commission or a person in charge of Education may also be a member. This board at municipal level, as well as the board at school level, prepares, applies and marks the test. In this case, they must draw up a proposal of 6 problems for the 3rd phase, which must be sent to the provincial board for approval. They are also responsible for disseminating the results and sending a report on the 2nd phase.

Meanwhile, the Provincial Commission or Tribunal shall perform the same functions as mentioned above, but shall be composed of the Head of the Provincial Department of Education, the Provincial Mathematics Coordinator, a representative

of the Commission of the Country in charge of Education and a Provincial Tribunal appointed by the Director of the Provincial Secretary of Education.
In the particular case of the National Selection Board, it is composed of the National Director of General Education, a specialist in Mathematics from the General Education Subsystem, a specialist from INADE (Instituto Nacional de Avaliagao e Desenvolvimento da Educagao) in the subject of Mathematics in the General Education Subsystem, a specialist from the Provincial Mathematics Centre of Luanda and a Provincial Mathematics Coordinator (on a rotating basis) who must maintain communication with the students of the team corresponding to the final phase of the competition.
This board is in charge of the elaboration of the Pre-Olympiad Mathematics test and of the technical indications for the supervision and implementation of the test in the 17 provinces except for the province organising the final phase. They have to select the 18 best results (6 for each of the 3 classes) and give the results.
This resolution specifies the functions of the presidents of the boards, for each of the levels and phases, as well as the criteria for the preparation and marking of the tests. It is on this basis that the winning students are selected to form part of the National Pre-selection of Mathematics. They are:
a) Use A4 size paper.
b) In legible handwriting.
c) Propose 5 (five) problems.
d) It must be resolved in 120 minutes.
e) The winner will be the one with the highest number of correct answers within the time limit.
f) In case there are students who have the same score, the exam should be reviewed again and priority is given to the one who demonstrates a remarkable use of logical thinking and innovative problem solving skills.
As can be seen, there are few criteria provided for the preparation and marking of the tests from which the most talented students in Mathematics are to be selected.
This document establishes the regulations for the development of the Mathematical Olympiad in Angola, where the process is described with the intention of participating in the OMCPLP. For this event, each member country develops national competitions after the preparation and pre-selection of the students. It is held annually and brings together, to compete in the solution of mathematical problems, four students of General Secondary Education under 18 years of age plus two teachers Kderes of these groups from each country.
Although Portugal and Brazil top the medal table for their undeniable development, Angola's results are very poor as they do not correspond to the efforts developed by the State in the Education sector. On the other hand, within Angola, the Huambo province is a province with a tradition of high cultural development in relation to the rest of the country, and has not been represented in the OMCPLP or in any other international Mathematics competition.
There are several reasons for these weak results in Angola and, in order to transform this reality, we can begin by analysing what countries that have good results in these competences are doing and apply these experiences in line with the Angolan context, in this respect Perez Almarales (2015) states that:

"According to (Smith, 2008) cited by (Navarro Cendejas, 2016), countries with success in international Olympiads have some of the following attributes: a large population, a significant proportion of their populations in good educational formulas, a well-organised preparation infrastructure to support mathematical competitions and a culture that values intellectual achievement. According to this author, the basis for talented students to develop is to have help in preparing individually for competitions, for which in some countries there are high quality books and a wealth of resources available on the Internet, including specialised Olympiad sites.

Obviously, Angola does not yet have the objective and subjective conditions according to the attributes proposed by Smith, although it has a population of more than 30 million inhabitants, there is a very low rate of secondary and higher education compared to the countries with success in the international Olympics as a result of colonialism and a war that devastated the country (Angola only achieved effective peace in 2002). So a preparation based on self-study with high quality books and abundant internet resources is still far from the Angolan reality, but one should not resign oneself to the results achieved.

Therefore, it is necessary to analyse what to do and to look for our own solutions in terms of the conceptions and the preparation model we have, with the resources available in the country, to use the knowledge competitions and the olympiad as a motivational element to improve the quality of teaching and to obtain better national and international results.

CHAPTER 3

The preparation of teachers for the development of mathematics competitions in the province of Huambo.

The Angolan government, through the Basic Law of the Education System of 2001, defined the general objectives of education as the harmonious development of the physical, intellectual, moral, civic, aesthetic and labour capacities of the young generation, in a continuous and systematic way, which allows them to raise their scientific, technical and technological level, with the aim of contributing to the socio-economic development of the country.

The preparation of teachers in Angola has been gaining significant spaces since the Education Reform in 2001, in order to achieve a gradual educational change, a quality education in the country for the implementation of the aspired social model.

In Angola, the process for teacher training is expressed in Law no. 17/16 of 7 October, which is the basic law of the Education and Teaching System. This system is structured in six Education Subsystems, where the Teacher Training Subsystem is the integrated and diversified set of bodies, institutions, provisions and resources destined to the preparation and qualification of teachers and other educational agents for all the other education subsystems.

The Teacher Training Subsystem is structured as follows:

a) Secondary Pedagogical Education.
b) Higher Pedagogical Education.

Secondary Pedagogical Education is the initial training process through which future teachers acquire and develop knowledge, habits, skills, abilities, capacities and attitudes that enable them to exercise the teaching profession in Pre-school Education, Primary Education and the I Cycle of Regular Secondary Education, Adult and Special Education and, according to criteria, access to Higher Education.

Secondary pedagogical education takes place after the end of the 9th grade and lasts four years in the Teacher Training Colleges for two-year courses of pedagogical professionalisation, depending on the speciality, for persons who have completed the 2nd cycle of secondary education. In-service teacher training is mainly provided by Teacher Training Centres or other educational institutions authorised for this purpose.

Higher Pedagogical Education is a set of processes, developed in Higher Education Institutions, aimed at training teachers and other educational agents, training them for the exercise of the activity and as support for teaching at all levels and educational subsistencies.

The Higher Pedagogical Education is studied after completion of the II Cycle of Secondary Education or equivalent, with a duration of four to six years depending on the particularities of the course, where it is necessary to acquire knowledge, skills, values and fundamental practices within a given branch of knowledge and subsequent professional training or academic training. Graduates of this sub-system are awarded the degree of Licenciado attained at the undergraduate level in accordance with Article 68 of the aforementioned law.

This subsystem is also responsible for postgraduate studies at two levels which confer the academic degree of Master or Doctor, according to Article 69 of

Postgraduate Studies. Higher Education Institutions may offer postdoctoral programmes, which do not confer an academic degree, and which aim at deepening, on the part of the candidates, the competences to carry out research.
The postgraduate course, which does not confer an academic degree, is aimed at the technical advancement of the person who has passed one of the undergraduate training levels and comprises:

a) Vocational training, with courses of variable duration.

b) Specialisation with courses of at least one year's duration, according to the areas of knowledge.

As can be seen, teacher training in Angola is legislated in the aforementioned law, which contemplates this process from a continuous training perspective, starting with initial undergraduate training, up to postgraduate training. However, the number of teachers without pedagogical training who are employed in the various subsystems is significant. The preparation of these teachers is a necessity and a challenge for the improvement of the Education and Teaching System.

Borges (2019) reaffirms that:

"The preparation needs of university teachers (graduates of non-pedagogical careers), who possess insufficiencies in pedagogical preparation and in particular in what refers to the training of skills from their psychological and pedagogical bases through a didactic adjusted to the demands imposed on their training".

An inescapable requirement for improving teaching in Angola is the professional preparation of teachers. However, the alternatives of preparation are only sometimes perceived, or restricted, to the national training plan whose proposal is framed in the academic training in master's and doctorate courses of Higher Pedagogical Education.

Several researches have been carried out and the results systematised by authors who have worked on issues related to the improvement of professional pedagogical performance in Angola (Cardoso, 2012), (Caimbo, 2013), (Da Costa & Kunjiquisse, 2012), (Liberato, 2014), which show the need for the preparation and improvement of teachers for these purposes and have allowed the identification of shortcomings for the development of professional pedagogical performance.

Among the problems mentioned about the preparation is that it is limited and unsystematic, the didactic and methodological elements are not intentionally taken into account and the content addressed is restricted, being poorly used to deal with the educational, social and economic problems that afflict the country.

No less distant from these problems is the preparation of secondary school mathematics teachers, who are responsible for the training of new generations with a high scientific level in accordance with the demands of the contemporary world.

One of their tasks is to meet the learning needs of their students individually, to motivate them in the study of mathematics and to participate in the mathematical olympiads. This is currently one of the challenges to be faced, as it demands a high level of professional performance, and shortcomings in this respect are evident from the initial training itself.

On the basis of the author's 12 years' experience as a teacher at the Teacher Training College, where he has been invited to give training seminars to secondary school teachers in various schools in Huambo, his exchanges with teachers and managers involved in the development of the Mathematics Olympiads, the review of the regulations and reports submitted by the Ministry of Education on the subject, it has been found that the following problems and shortcomings in the performance of teachers in terms of the following are evident:

- The identification of the gifted student in Mathematics from the teaching-learning process of the subject in secondary schools.
- Motivation of Secondary School students to study Mathematics and to participate in competitions.
- The lack of teachers responsible for the preparation and care of gifted pupils in secondary schools.
- The school organisation does not envisage differentiated attention for gifted pupils for their participation in the mathematics competition in secondary schools.
- Teachers do not have the undergraduate or postgraduate academic preparation for differentiated attention to gifted pupils.
- Self-preparation is the main form of teacher preparation for differentiated attention to gifted pupils.
- The preparation of the teachers for the training of the competitions is included in the preparation.

In addition to all of the above, there are limitations in terms of the preparation and projection of postgraduates in terms of the problem posed, and at the same time the methodological limitations faced by mathematics teachers in attending the competitive examinations are not directly addressed.

With regard to the preparation of teachers for the care of student competitors, Perez (2015) states that:

Teachers need to be prepared to teach the content of mathematics competitions, as they did not receive this in their training and the number of postgraduate courses in this area is limited. On the other hand, they do not have the necessary bibliographies for their preparation, and there is no syllabus to guide them in their preparation.

It then points out the need to

The teacher-preparer prepares the teachers of the schools in the context, from a teaching and methodological point of view, which includes determining how to work towards the achievement of each objective of the programme. (Perez, 2015).

In the article "Capacitacion del profesor que entrena para los concursos de matematica en la educacion media" its authors make important remarks regarding the scientific-methodological deficiencies of the teachers to develop an effective work as trainers of contestants in the province of Matanzas, Cuba, which, after the verification that they exist in the educational context of Angola, have been taken by the author as guidelines to structure the preparation of Mathematics teachers in the activity of competitions in the province of Huambo:

> In the training of mathematics teachers, it is necessary to reinforce the work on competitive training and on the teachers working in different educational institutions through: training seminars, methodological preparation, individual improvement and different forms of postgraduate education.

> Knowledge of heuristic problem-solving procedures and strategies is limited, leading to non-achievement of the objectives at the level with respect to independent understanding of mathematical problems, and insufficient use of graphical models to

support text comprehension.
> It is necessary to prepare teachers on how to use general and particular heuristic strategies, when known general or particular principles are not effective.
> The methodological preparation of mathematics teachers in schools must meet the needs of each teacher to prepare them for their professional performance in the classroom.
> For the development of methodological work, different ways can be used, depending on whether individual or collective forms are used. Methodological preparation activities should be dominated by demonstration, modelling (with possibilities for discussion) and reflection, all skills which foster creativity and professional development of teachers.
> Another variant used in the permanent preparation of mathematics teachers is the preparation seminars on the contents of a grade at the beginning of each school year. The essential objective of these seminars is to raise the level of preparation of teachers, considering the real conditions of the diagnosis of teachers in the territory, the design of this preparation responds to objective needs of the aspects of content and teaching strategies that should be used so that students learn more in each grade.
> Other preparatory activities should be organised in response to specific educational development needs, such as: self-preparation, specialised conference, seminar, workshop, scientific debate and others, including methodological work.

In order for Mathematics teachers to be successful in training Secondary School pupils participating in Mathematics competitions, they need to have mastered the mathematical content and the strategies and techniques required to manage this process of thinking development, so that they achieve independence in solving arithmetical problems.

In this regard, they point out that:

At the same time, actions aimed at the permanent training of mathematics teachers should be developed, taking advantage of the spaces established for this purpose: methodological preparation in each centre according to the needs and potential of the teacher, tasks aimed at individual self-preparation, concentrated teacher preparation seminars and the different modalities of postgraduate education (Carazo et al, 2016).

In the particular case of the province of Huambo, the group of teachers who teach Mathematics in Secondary Education is numerous. On the other hand, this group is heterogeneous in terms of their training, where a significant part has no pedagogical or didactic training in Mathematics, and where some have a university education and others do not.

In this regard, it should be noted that before the publication of Presidential Decree no.160/18 of 3 June 2018 (Estatuto de la Carrera de los Agentes de la Educacion), the Angolan educational model allowed for the entry of professionals in the education sector. Although in some cases these teachers were given a short training in some didactic subjects before joining, it is not enough to develop the teaching work with the desired efficiency.

These factors conspire, from the legislation of the Angolan Educational System, the conception of a plan of preparation and professional training or specialisation, since according to the regulations only graduates of any Higher Education degree can attend these courses.

This has led to particular views on the ways of selecting contestants, on the structure and design of the tests, and different evaluation criteria for talented pupils. As a result, the contestants are poorly prepared for the tests at the final stage of the competitions, let alone at the international level where the demands on mathematical skills are higher.

Other factors negatively affecting Huambo's poor performance at the national level include the following:

- The preparation of students for participation in mathematics competitions in the schools of Huambo is not promoted or stimulated at school level.
- In the different stages of the first phase of the competition, no reports or rapporteurships are drawn up as indicated for the two higher phases.
- In the other phases, reports or reports with administrative characteristics are produced, without mentioning the main achievements and difficulties of the contestants in solving the problems. This means that it is not possible to follow up the contestant in the next stage or phase, on the basis of the mistakes made, i.e. a diagnosis of this and of the process itself, as well as of those contents most affected in the Education System.
- The publication of test syllabuses is not authorised. This means that the trainers or coaches of contestants do not have the resources to improve their work, nor do they have control over the evolution of a process designed for educational purposes, as well as limiting the researchers who could contribute to its enrichment.

Another of the contradictions to be faced is given by the syllabus of the discipline of Mathematics itself, which does not have an adequate systematisation and preparation of teachers. Likewise, regarding the dilemma of mathematics education in Angola, (Gangula & Faustino, 2018) mentions the insufficient didactic, pedagogical and technological preparation of teachers, and the decontextualisation of curricula and lesson plans as the core problems to be solved if superior results are to be achieved in the educational sphere in general.

On the other hand, most of the teachers make a daily journey of more than 60 km to reach their jobs, as the vast majority of the teaching staff live in the municipality of Huambo (the main urban centre of the province). This situation is aggravated by a public transport network that is deficient in some municipalities and non-existent in others, which is economically, physically and morally exhausting.

In addition, there is no promotion of high mathematical competencies for the positions of mathematics co-ordinators in the different schools in Huambo. This situation extends to other curricular disciplines. It is difficult to mention, but it is real. Nepotism, cronyism and aspects of poritic link to the dominant party are relevant. However, in the year 2021, Presidential Decree No. 93/21 of 16 April was published, which establishes the necessary profile for the exercise of leadership positions in public educational institutions, where it is regulated that coordinators must have specific training in the discipline, course or area they are going to coordinate and positive performance evaluation in the last five years.

Regarding this, in their work (Cameira, Rodriguez and Garrta, 2018, p.72) they point out that "one of the great evils of the teaching centres is that we do not have at the top people trained to exercise the real role of pedagogues, transformers of consciences, and capable of changing the teaching practices of their teachers". In the same study, these authors concluded that 50% of the managers are not trained in educational sciences, they are appointed for convenience or considering the long time of service as a teacher, which is detrimental to the teaching work.

This perspective favours the exclusion of the contributions of teachers with high mathematical competences in the process of improving the teaching-learning process in this discipline. Despite the fact that in 2017 policy reforms were initiated where, among several goals, they suggest a more inclusive participation of human resources for the sustainable development of the country, and in 2021 the Decree that establishes the necessary profile for the exercise of leadership positions in public education institutions was published, significant changes in this regard have not yet been registered.

On the other hand, the national context still does not favour education with the desired quality. It is enough to see that between 2010 and 2015 less than 9% of the General State Budget was invested in the education sector, according to official documents of the government of the Republic of Angola. On the other hand, the joint reports for the years 2016 to 2022 of the non-governmental organisations UNICEF, Action for Rural and Environmental Development (ADRA), and the Angola Social and Political Observatory (OPSA), show that Angola has invested less than 7% of its general budget in the education sector during this period, which translates into the lowest rate compared to the other countries in the Southern African region.

One of the consequences of the above is the lack of teaching facilities and the limited amount of physical space for teaching. On average a Primary and Secondary teacher works with groups of 60 students.

In the National Development Plan 2018 - 2022, it is highlighted that the secondary education subsystem faces several challenges, such as the insufficient number of classrooms and teachers to meet the demand for education, as well as the existence of a precarious infrastructure. Despite the fact that this document has established as one of the priority actions the construction of more schools and classrooms with emphasis on the most deprived regions, this shortage has not yet been minimised. To this must be added the fact that Huambo province is part of this reality.

In this regard, in their reports from 2016 to 2022, the philanthropic organisations ADRA, OPSA and UNICEF stress that there are still some inconsistencies between the government's political priorities in the National Development Plan and the financial allocation in the General State Budgets from 2013 to 2022.

From the point of view of (Rodriguez and Hinojo, 2017) cited by (Cameira, Rodriguez and Garrta, 2018, p.72), "improving the quality of education goes through several stages, such as the construction of new school structures, the equipment that schools must possess in correspondence to their level of teaching, the monitoring of these facilities and, above all, trained personnel to face social demands".

What is clear from the above is that it is necessary to prepare the teachers for the development of the mathematics competition in the province of Huambo:

"Even without taking into account the aspects related to economic and social processes that deeply disturb the teacher's condition and performance, many problems more directly linked to their training in the scientific and pedagogical components, compromise the current model of teacher education, making it unproductive and disconnected in the achievement of its objectives". (Juliao, 2020, p.10)

In order to respond to the need to prepare teachers methodologically for the development of the Mathematics competitions, in April 2018 the Provincial Directorate of Education in Huambo entrusted the task of giving training seminars to Mathematics teachers, which was entrusted to teachers Alambre Jose Pinto and Jose Chiumbo Paiva.

On the basis of a diagnosis made to teachers by the author, the study of the mathematical contents required in Mathematics competitions and the didactic and methodological demands for their treatment, the following topics were selected in this order:

1. Organisation of the Mathematical Olympiads in the light of their regulations.
2. Mathematics competitions in Huambo province: a pedagogical model for their development.
3. Problem teaching, its objectives and methods in the teaching-learning process. With examples of topics related to Mathematics Competitions and Olympiads for secondary education.
4. The methodological treatment of mathematical exercises and problems. The general heunstical programme, its usefulness for the conscious and active work of the students in the solution of exercises and problems. Emphasis is given to the topics related to mathematical competitions, which are:

- Flat geometry,
- Teona of numbers,
- Combinatorial calculus,
- Financial mathematics,
- Techniques of demonstrations of mathematical propositions, and
- Mathematical situations whose solutions do not correspond completely to the contents studied by the student, but with a reasonable level of complexity so that they are within their reach.

A total of 35 teachers participated in these events; 2 pedagogical supervisors of the Provincial Directorate of Education and the remaining 33 are the coordinators of the discipline of Mathematics of the Zones of Pedagogical Influence (ZIP), and the coordinators of the discipline of Mathematics of the I and II Cycles of Secondary Education in each municipality.

One of the purposes was for these coordinators to serve as trainers of teachers and contestants in their function, since a ZIP is a network of primary schools that are part of a municipality and are coordinated from a reference school in the district.

The first seminar took place in the second half of May 2018, taking advantage of the students' pedagogical break at the end of the first term of the school calendar. The other seminar with the same workshops except for the pedagogical model, but with a greater degree of depth, was held in the second edition of the seminars in the same period.

Each of these events took place on 5 working days of the week, with a duration of 40 hours, 36 hours of which were dedicated to teaching and 4 hours to the opening and closing sessions.

Presentations of the workshops and the results of the teacher preparation seminars

Theme 1: *Organisation of the Mathematical Olympiads in* the light *of their regulations*, enabled the participants in the workshops to understand these regulations, which is one of the factors in developing the competitions with the required systematicity.

In topic 2: *Mathematics competitions in Huambo province: a pedagogical model for their development*, the participants in the workshops will study the most relevant components of the teachers' preparation requirements for the development of the Mathematics competition, which are detailed in the following section.

The presentation and discussion of the workshops in theme 3, concerning problem teaching, its objectives and methods in the teaching-learning process, developed in the participants the skills for the conception of efficient teaching methods in the conduct of training of the contesting students and in general in the development of a mathematics class.

In this regard (Haydt, 2011, p. 214) points out that the problem-based teaching method allows:

- "Stimulate students' participation in the construction of knowledge, triggering their mental activity by mobilising their operative schemes of thought.
- To develop reasoning and reflection.
- To favour the acquisition of knowledge, enabling its application in practical problem-solving situations.
- Facilitate the transfer of learning by applying knowledge to new situations.
- Develop initiative in the search for new knowledge, decision making and problem solving".

In the presentation and discussion related to *the methodological treatment of mathematical exercises and problems, and the general heunstical programme*, the participants became familiar with typical problems of mathematical competitions, as well as appropriated knowledge that sharpens the ability to solve exercises and problems where it requires the productive or creative application of the preceding mathematical knowledge or their creative ingenuity to the solution of a problem situation that is new to them. Founding this:

"The teacher for high abilities should have some of the same characteristics and abilities of their gifted students and in this line indicated that teachers working with students with special talents should also have special skills and knowledge regarding the particular characteristics of these children that facilitate their personal, social and academic development". (Feldhusen, 1997) cited by (Conejeros-Solar, Gomez-Arizaga and Osorio, 2011, p.397).

During the closing talks of the training seminars, the teachers expressed their satisfaction both with the initiative of the Provincial Directorate of Education and with the themes developed in these events. They were very grateful and at the same time pleased with the double orientation of the day, i.e. the possibility to have access to lectures and also to participate with contributions to improve the performance of our contestants in the Mathematics Olympiads and of the teaching work in general.

[a]It should be underlined that one of the achievements of these training seminars was the winning of a gold medal by a 9th grade contestant from Huambo province in the

8th edition of the national competition "Olimpiadas de Matematica" (Mathematics Olympics).

CHAPTER 4

The components of teacher preparation requirements for the development of Mathematics Competitions

As a possible solution to the problems in the preparation of teachers for the development of mathematics competitions (DCM) in Huambo province, it was necessary to determine the requirements for teacher preparation (EPD).

The most relevant components of the EPD are:

1. The identification of gifted students in the Mathematics Olympiad.
2. The typology of the problems in the Mathematics Olympiad.
3. The structure of the tests in the Mathematics Olympiad.
4. Measurement criteria for the identification of gifted students in the Mathematics Olympiad.
5. The reports as a result of the Mathematics Olympiads.
6. Differentiated educational attention and the preparation of gifted pupils for the Mathematics Olympiad.

The following flow chart shows the relationship between the 6 elements mentioned above and which form the basis of the proposed EPD.

Components of the EPD for the DCM.

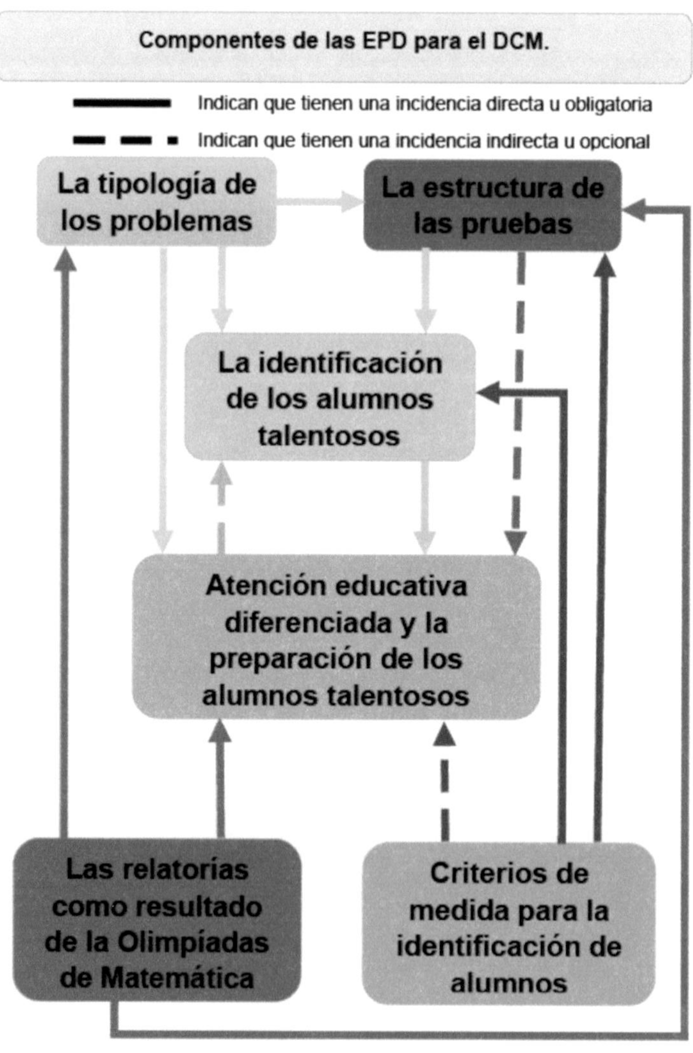

Indicate that they have a direct or obligatory impact
Indicate that they have an indirect or optional impact
The typology of the problems The structure of the tests
Identifying talented pupils
Differentiated educational care and the preparation of gifted pupils
The reports as a result of the Mathematics Olympiads
Measurement criteria for the identification of pupils

1. Identifying gifted students in Mathematics

Correctly identifying talented students is not an easy task, even if it is intended to be achieved as early as the Olympiad tests, which start at the beginning of the school year, where teachers may not know their students well in the first place.

The characteristic of mathematically gifted subjects according to (Werdelin, 1958) cited (Susana & Jhonny, 2014).

"...is the ability to understand the nature of mathematics, problems, symbols, methods and rules; the ability to learn them, retain them in memory and reproduce them in combination with other problems, symbols, methods and rules; and the competence to use them in solving mathematical tasks".

In terms of methodology, the identification of talented students in the Mathematical Olympiad is based primarily on performance in competitions and tests, where the application of mathematical skills, problem solving and logical thinking are assessed. In addition, the results can be considered together with the history of participation in previous competitions and other relevant factors to determine a student's mathematical talent.

It is important to note that the theoretical and methodological underpinnings may vary according to the country and the policies of the Mathematical Olympiad. Different organisations and competitions may have slightly different approaches to the identification and selection of talented students. But in general, the identification of gifted students in the Mathematical Olympiad is based on a theoretical and methodological foundation. Some of them are mentioned below:

S *Exceptional performance in Mathematics*: Gifted students in the Math Olympiad generally perform exceptionally well in the subject. They have a broad understanding of mathematical concepts and demonstrate superior skills in solving challenging problems.

S *Logical thinking and mathematical creativity*: Gifted students show advanced logical thinking and are able to apply mathematical concepts creatively to solve complex problems. They have a solid understanding of mathematical structures and patterns, and are able to make connections between different areas of mathematics.

S *Ability to solve additional problems*: Gifted students in the Mathematical Olympiad are able to solve additional problems beyond the school curriculum. These problems often require deeper logical reasoning, greater ability to abstract and generalise, as well as unconventional approaches to problem solving.

S *Teamwork skills*: Although the Mathematical Olympiad is primarily an individual competition, talented students must also demonstrate teamwork skills in certain contexts, such as in team relay events. Effective collaboration, clear mathematical communication and the ability to share ideas are valued in these scenarios. This ability has not been considered in our work, but may be considered in the future to be in line with international trends.

In general, when referring to mathematical talent, emphasis is placed on the ability to apply their knowledge and the mathematical procedures they have received in general education in a creative way (Palacios, 2006).

Therefore, the success or failure of the process of identifying gifted students in Mathematics from the Olympics will depend fundamentally on the characteristics of

the test applied to the students. In this regard (Prieto et al, 2002) state:
"Assessment environments must meet a number of requirements: integrating curricular content and assessment materials designed to assess competence in the different intelligences (knowledge, skills, attitudes and working styles).
From what has been said, it can be inferred that the typology of problems constitutes an essential resource for the identification of talented pupils on the basis of solving them, both in their preparation process and in the olympiad tests according to each of the competitive levels. For their study, and with the intention of including problems that contemplate both the curricular mathematical content and that value and stimulate creativity, the classification of problems proposed for the EPD is set out in the following epigraph.

2. The typology of problems in Mathematics Competitions

When we speak of mathematical problems, we do not refer to the trivialised version of text exercises, also known as "word problems" in English. What is a mathematical problem?
There are various definitions available in the literature; many of them assume views of what a problem is, which create a harsh flavour. They convey the feeling that a problem is something like a trap. For example, for (Cruz, 2006) cited by (Nieto, Lizarazo & Carrasco, 2015) the term problem refers to truly complex situations, capable of enhancing the development of students' thinking, and of providing ways of acting to face the challenges of science, technology and everyday life. Such situations^ are difficult to find in educational practice.
Another author who tried to define the term mathematical problem is (Miguez, 2002). He states in his work that "a mathematical problem is a real or fictitious situation that challenges the conceptual understanding, and not only the knowledge of a subject dealt with in the mathematics learning activity; it demands a restructuring in the way of approaching the situation posed and the limits of the known procedures, and seeks to generate connections on varied knowledge". For this author, a problem has no time limit, it can be solved quickly, or it can never be solved at all.
According to (Nieto, Lizarazo & Carrasco, 2015), who makes an interesting review of authors who have tried to define and characterise the term problem is (Pino, 2013). This author presents a table that summarises different contributions on the characteristics that a mathematical activity must have in order to be considered a problem, and in which the following stand out:
S The need to have a goal that we cannot easily reach with an immediate process;
S Doubts and/or blockages generated by the situation raised or by the lack of a clear method that leads to the solution;
S Accepting the challenge consciously in order to reach what may be considered by the resolver as a personal challenge, and
S The use of mathematical concepts and processes.
The approach to problems for the mathematical olympiads should be structured in such a way that the pupils can acquire a more refined knowledge of mathematics, allowing them to glimpse a promising prospect of a possible correct solution through formulations or conjectures. Also, these problems should be such that they can lead to the student's understanding of all stages of the solution.

How to classify mathematical problems? Numerous authors have proposed classifications of mathematical activities and problems. Starting with G. Polya, in 1945, with his well-known book "How to solve it", we can go on citing (Ballester et al, 1992), (Blanco, 2001) and (Angel M^guez, 2002) who classify mathematical problems according to different criteria.

Polya's classification takes up the distinction made by the Greeks, between theorem and problem. This mathematician mentions only two types of problems: "problem to be solved" and "problem to be proved", the criterion of distinction refers to the objective of the problem. The parts that constitute the problem are different depending on the case, in a problem to be solved there are unknown, data and condition, and in a problem to be proved there are hypothesis and conclusion.

Most of the authors consulted make this type of classification, considering only two categories. For example, (Niss, 1991) classifies mathematical problems into applied and pure problems, (Noda, 2001) and (Santos, 2014) classify them into "well-structured" and "ill-structured" problems.

This last typology refers to the structure and formulation of the problem. For these authors, well-structured problems are those that we find in school textbooks where all the information necessary to solve them is contained in the statement, the rules for finding the correct solution are clear and there are defined criteria for verifying the solution. Poorly structured problems are those that do not present a well-defined structure, and where the information may be insufficient or excessive, which requires them to be reformulated, in addition to the fact that their solution requires different processes and criteria to be considered.

Another classification based on two categories is by Pehkonen (1997) cited by (Nieto, Lizarazo & Carrasco, 2015). This author classifies mathematical problems into open and closed problems. This distinction refers to the degree of accuracy of the description of the starting and ending situations. In a closed problem the beginning and the end are exactly explained in the task. If either the starting or the end situation is open, then we have an open problem and they are characterised by the possibility of having different solution strategies.

However, we have to consider that no classification can be exhaustive, as there are always intersections between the different sections and there are activities that are difficult to catalogue, given the diversity of problems that can be proposed at different levels and with different contents.

It is known that mathematics is a deductive science, which deals with the study of the properties of abstract entities, as well as the connections and relationships that exist between them. Therefore, "mathematical reasoning is par excellence hypothetico-deductive reasoning, although inductive reasoning also plays a fundamental role in mathematical activity, since it presides over the formulation of conjectures" (Programa de Matematica A - Ensino Secundário de Portugal, 2013).

On the basis of the above, a first classification of the mathematics Olympiad problems into two types is proposed for their study:

The Olympic problem eminently deductive (POED): one whose solution is constructed from the combination of mathematical properties already taught to the student from the primary level to the level at which he/she is, that is, in this problem the student must apply in a productive or creative way the previous mathematical

knowledge to the solution of a problem situation that is new to him/her. Examples:
Plane or spatial geometry problems that combine different figures into a single figure.
Algebra problems combining equations of different degrees.
Demonstrations of non-routine mathematical properties from known properties.
Olympic problem eminently inductive-deductive (POEID): is one whose solution path does not correspond entirely to the contents studied by the student, but with a reasonable level of complexity so that it is within his reach; in this case the student must apply not only the previous mathematical knowledge, but also his creative ingenuity, reaching the so-called lateral thinking or solutions by means of non-conventional strategies or algorithms. Examples:

- Problems that lead the student to the solution of problem situations that constitute propositions or properties studied in a specific branch of mathematics.
- Problems whose solution induces the student to generalise by means of the construction of mathematical properties that encourage their demonstration.

It is necessary to specify that, for both types of problems proposed, they must be interesting, original, which cannot be very similar to those that the competitor may or may not have solved in his preparation, so that he does not have a priori an idea of a solution.
One of the functions of the eminently deductive Olympic problems is to diagnose the process of teaching and learning (PEA) of mathematics at each level of schooling. Beyond detecting the gifted pupil, attention should be paid to the intra-school content in order to detect the main difficulties faced at that level, which will allow more appropriate didactic actions to be devised.
On the other hand, this typology of problems makes it possible to verify the solidity of the appropriation of knowledge on the part of the students, as well as to encourage teachers to comply with the programmes and syllabuses of the discipline of Mathematics conceived by the Ministry of Education.
It is also important to emphasise that they contribute to the development of students' logical and abstract thinking, their critical and creative capacity, through the assimilation of mathematical concepts and methods, theorems and their proofs.
In addition to helping in the diagnosis of the degree of solidity of the knowledge acquired by the pupils up to the level where they are, the inductive-deductive type of empirical problems stimulate their creativity and independent spirit, as well as directing them towards greater cognitive independence.
This type of problem is typical of the international Mathematics Olympiad tests, allowing the identification of talents as usual, but does not always guarantee the motivation of students to study this discipline, since the solution of such exercises, in many cases, requires more ingenuity than deep knowledge, which has the risk of inducing the student to think that the mathematics of the Olympiad is different from that taught in schools.
The use of this classification serves several purposes:
1. *Description and communication:* Classification allows Olympic organisers and participants to have a clear understanding of what kind of problems they can expect in the competition. It helps to communicate and describe the problems more accurately.

2. *Diversity and balance:* The classification ensures that there is a variety of problems in the competition, from more rigorous problems focused on logical deduction and application of acquired knowledge to more exploratory problems requiring inductive-deductive reasoning. This allows different problem-solving skills and approaches to problem solving to be assessed.
3. *Development of mathematical skills:* Sorting allows participants to develop and improve their expected skills of reasoning and applying what they have learned (logico-deductive) and inductive-deductive reasoning. By facing problems of both types, students can strengthen different aspects of their mathematical thinking.

The advantages of this classification include:

- It provides a clear and structured description of the problems for participants and organisers of the Olympics.
- It allows a more balanced assessment of the participants' mathematical skills.
- Encourages the development of various mathematical reasoning skills.

Disadvantages may include:

- Classification can be somewhat subjective and problems can have both deductive and inductive-deductive elements.
- Some participants may have a preference or strength towards one of the problem types, which could influence their results.
- It can limit diversity in problem choice and design.

The following diagram summarises what has been said about the characterisation of the typology of the problems proposed for the Mathematics Olympiad tests.

Characterisation of the problem types for the Mathematics Olympiad tests

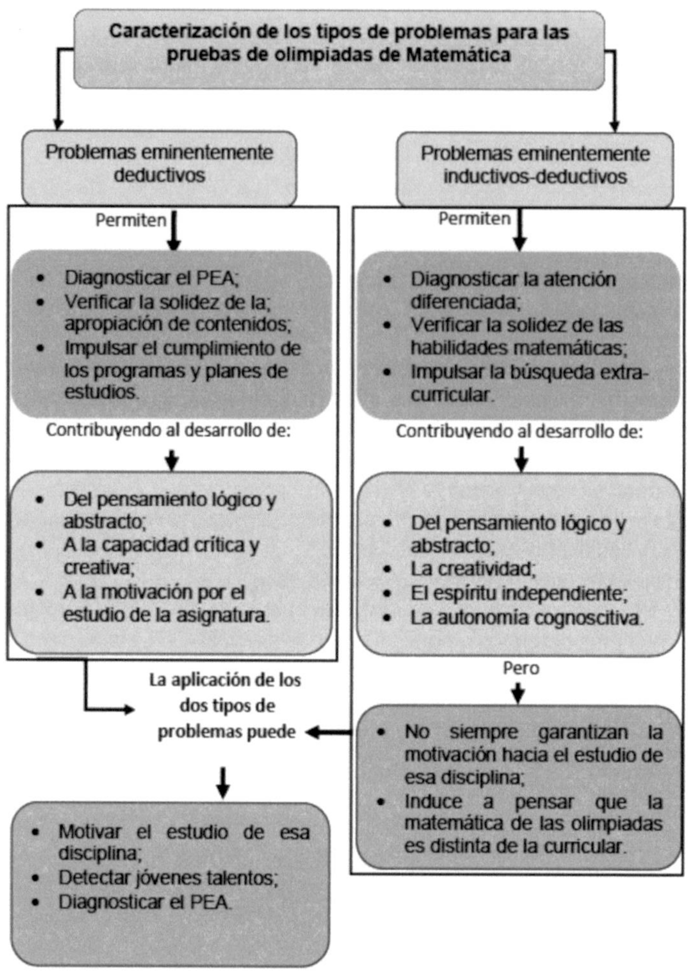

Eminently deductive problems
Eminently inductive-deductive problems
Allow Allow
"Motivate the study of this discipline; ∀ Detect young talents; ∀ Diagnose the EAP.
"It does not always guarantee motivation to study the discipline; ∀ It induces people to think that the mathematics of the Olympiads is different from that of the curriculum.
The application of the two types of problems can
But
"Logical and abstract thinking; " Creativity; ∀ Independent spirit; " Cognitive autonomy.
Contributing to the development of:
"Diagnose differentiated attention; ∀ Verify the strength of mathematical skills; ∀ Encourage extra-curricular research.

Contributing to the development of:
"Diagnose the ASP; ∀ Verify the soundness of the; appropriation of content;
∀ Promote compliance with programmes and curricula.
"To logical and abstract thinking; ∀ To critical and creative capacity; ∀ To motivation for the study of the subject.

3. The structure of the tests in Mathematics competitions

Under the conception of these SPU, the Olympiad tests should not be seen as mere instruments to measure the mathematical competences of the contestants; it is suggested to perceive it as an evaluation process. In this respect (Prieto et al, 2002) point out that:

"Assessment should be process-oriented rather than product-oriented, as it allows us to obtain valuable information from the learner as he/she carries out an activity within the curricular context. The individual profile of the learner's intelligences should be drawn in order to detect his or her skills as well as possible gaps or deficiencies".

In their work (Michailuk & Nicodemo, 2015) they state that if we focus specifically on exams or tests as assessment instruments, we can make two types of analysis. One by considering the characteristics of the test independent of the particular resolution by a student: the number of problems it contains, the concatenation of items, the length, etc. The second is by looking at post-implementation issues, linked to correction and evaluation: the type of answers allowed, how these different answers are graded, etc.

These authors also point out that when planning an evaluation or designing evaluation instruments, the following aspects should be considered in their construction:

- The distribution of contents and tasks,
- The sequencing of problems,
- The assessment of responses and,
- Correction and feedback.

This line of thought leads to the adoption of a test structure for each stage of the Olympiad, so that in the process of carrying out the test, information is obtained on the potential and weaknesses of the students, with a view to contributing to the improvement of the Mathematics ASP and the Olympiad itself.

In this respect, it is worth reiterating that the number of problems for the Olympiad test and the time allowed to solve them are set out in paragraphs b), c), d), e) and f) of Article 20 of the Regulations of the Olympiad "Olimpiadas de Matematica". Nevertheless, a structure of the test is proposed here which, from the didactic point of view, helps to achieve the objectives of the Olympiad itself, without affecting that set out in article 20°.

In the following, a structure for the tests is proposed in terms of the dosage of the types of Olympic problems:

- For the intra-group student olympiads: 80% of eminently deductive oKmpic problems and 20% of eminently inductive-deductive oKmpic problem.
- For the inter-group Olympiad: 80% of eminently deductive oKmpic problems and

20% of eminently inductive-deductive oKmpic problems.

- For the inter-school Olympiad (municipal): 60% of eminently deductive Olympic problems and 40% of the eminently inductive-deductive type.
- For the inter-municipal (provincial) Olympiad: 40% of eminently deductive Olympic problems and 60% of eminently inductive-deductive type.

The smaller proportion of eminently inductive-deductive oKmpic problems is so as not to run the risk of what has already been said about inducing the pupil to think that the mathematics of the olympiads is different from that taught in schools, since this type of problem requires more ingenuity than profound mathematical knowledge.

On the other hand, the higher proportion of eminently deductive oKmpic problems can help in the diagnosis of the Mathematics ASP, detect talented students and serve as a basis for the preparation of the Olympiads where the demand generally falls on the eminently inductive-deductive oKmpic problems, because, although this type of problems do not require high knowledge of mathematics, they demand solid bases even at the level at which the Olympiad is held.

It is also considered that the combination of the two types of oKmpic problems (eminently deductive and eminently inductive-deductive), besides helping in the diagnosis of the degree of solidity of the knowledge acquired by the students up to the level where they are, stimulates in them creativity and independent spirit, as well as directs them to a greater cognitive autonomy.

As can be seen, the structure of the tests for the "Olympiads of Mathematics" has a gradual and progressive character by defining for each of the levels the quantities that must be present according to the typology of the problems proposed. This can help in the evaluation of the PEA of Mathematics in the Angolan context, detect talented students and serve as a basis for the preparation of the international Olympiads.

3.1. Examples of tests from POED and POEID mathematics competitions.

In this section we present examples of two tests for the eleventh grade "Olympiads in Mathematics", which combine in their structure mainly deductive oKmpic problems (POED) and mainly inductive-deductive oKmpic problems (POEID), and gradation of the complexity of the same in the two phases of the competition.

TEST 1: Intra-group or inter-group phase

1. Suppose that $2023 = (n-2)^n(n+1)^{n-1} + 23$. Determines the value of $n, (n \in \mathbb{Z}^+)$.

2. Consider a nine-sided convex poKgon, in which the measures of its internal angles constitute an arithmetical progression of reason equal to 5°.
What is the greatest angle of the poKgon?

3. Show that a natural number ***n*** of three hundred digits, consisting of one hundred digits 0, one hundred digits 1 and one hundred digits 2, can never be a prime number.

4. The numbers

$A = \sqrt{4+\frac{2}{3}\sqrt{7}} + \sqrt{4-\frac{2}{3}\sqrt{7}}$ y $B = \sqrt{57-40\sqrt{2}} - \sqrt{57+40\sqrt{2}}$ are integers. Determine $(A+B)^2$.

5. (OBMEP, 2017) In the following sum each letter represents a number. Determine the value of WATER.

DROP + DROP + DROP + DROP + DROP = WATER

In this test, the first four exercises can be considered as POEDs and the last one can be considered as a POEID.

To solve this exercise we need to decompose the number 2023 into product and sum at the same time, and with some creative ingenuity achieve the value of n, as follows:

$$2023 = 2000 + 23$$

Decomposing 2000 into prime factors we obtain that $2000 = 2^4 \times 5^3$.

Evidencing creative ingenuity, it can be seen that

$$3 = 4 - 1$$

$$2 = 4 - 2$$

$$5 = 4 + 1$$

where it can be "discovered" that $n = 4$.

Let's see a way to solve *exercise* 2. In this way it is necessary that the contestant masters some properties of convex poKgons and arithmetical progressions, as well as the methods to solve systems of linear equations of the type 2 x 2.

In the syllabus of the Mathematics discipline in Angola, the subject of *arithmetical progressions* is dealt with in the eleventh grade, and the other mathematical situations required to solve the exercise, as we shall see below, are studied at the appropriate levels. Let us see:

Let us consider $A = \{a_1, a_2, a_3, \dots, a_9\}$ *an arithmetic progression (AP) of reason* $r = 5^o$.

$$a_n = a_1 + (n-1)r \quad \Rightarrow a_9 = a_1 + (9-1) \cdot 5^o \Rightarrow a_9 = a_1 + 40^o \qquad (1)$$

The sum of the internal angles of a convex polygon is given by $S = (n - 2) - 180°$, *where* n *is the number of sides. Then the sum* S_9 *of the internal angles of a 9-sided convex polygon is*

$$S_9 = (9-2) \cdot 180^o \quad \Rightarrow S_9 = 1260^o \qquad (2)$$

The formula for calculating the sum of the n *therms of a PA is*

$$S_n = \frac{(a_1 + a_n) \cdot n}{2}$$

Hence

$$S_9 = \frac{(a_1 + a_9) \cdot 9}{2} \qquad (3)$$

Substituting (1) and (2) into (3), we get

$$1260^o = \frac{(a_1 + a_9) \cdot 9}{2} \qquad \Rightarrow a_1 + a_9 = 280^o \quad (4)$$

Solving the system formed by equations (1) and (4) ,

$$\begin{cases} a_9 = a_1 + 40^o \\ a_1 + a_9 = 280^o \end{cases}$$ *you get* $a_1 = 120^o$ *y* $a_9 = 160^o$.

Solution: The largest angle is $160°$.

As can be seen, the solution was constructed from the combination of mathematical properties already taught to the student from the primary level to the level he is at now.

In the case of *exercise 3*, which can also be considered POED, although at Secondary School level the methodological treatment of the *techniques of theorem demonstrations* is not done, the pupil has already been taught several theorems or elementary mathematical properties and their demonstrations up to the eleventh grade. On the other hand, the concept of prime number and the criteria of divisibility in the set $T+$ of positive integers is a subject already studied in the sixth grade.

In order to achieve the solution of this exercise, the contestant can productively or creatively apply the preceding mathematical knowledge in the following way:

The sum of all the digits of the three-hundred-digit natural number n *made up of 100 zeros, 100 ones and 100 twos is*

$$S = 100 + 200 = 300$$

which is a multiple of 3. Therefore, however we form the number n*, it will be a multiple of 3 and, therefore, it will not be a prime number.*

As for *exercise 4,* it can be solved as follows:

$$A = \sqrt{4 + \frac{3}{2}\sqrt{7}} + \sqrt{4 - \frac{3}{2}\sqrt{7}} \Rightarrow A^2 = \left(\sqrt{4 + \frac{3}{2}\sqrt{7}} + \sqrt{4 - \frac{3}{2}\sqrt{7}}\right)^2$$

$$A^2 = 4 + \frac{3}{2}\sqrt{7} + 2\sqrt{\left(4 + \frac{3}{2}\sqrt{7}\right)\left(4 - \frac{3}{2}\sqrt{7}\right)} + 4 - \frac{3}{2}\sqrt{7}$$

$$A^2 = 8 + 2\sqrt{16 - \frac{9 \times 7}{4}} \Rightarrow A^2 = 8 + 2\sqrt{\frac{64 - 63}{4}}$$

$$\Rightarrow A^2 = 8 + 2\sqrt{\frac{1}{4}} \Rightarrow A^2 = 9 \Rightarrow A = -3 \text{ ó } A = 3$$

The following analysis: $\sqrt{4 + \frac{3}{2}\sqrt{7}} > 0$ y $\sqrt{4 - \frac{3}{2}\sqrt{7}} > 0$, leads to the conclusion that

$$\sqrt{4 + \frac{3}{2}\sqrt{7}} + \sqrt{4 - \frac{3}{2}\sqrt{7}} > 0,$$

therefore, $A = 3$.

$$B=\sqrt{57-40\sqrt{2}}-\sqrt{57+40\sqrt{2}} \Rightarrow B^2=\left(\sqrt{57-40\sqrt{2}}-\sqrt{57+40\sqrt{2}}\right)^2$$

$$B^2=57-40\sqrt{2}-2\sqrt{(57-40\sqrt{2})(57+40\sqrt{2})}+57+40\sqrt{2}$$

$$B^2=114-2\sqrt{57^2-40^2\times 2}$$

$$B^2=114-2\sqrt{49}=100$$

$$B=-10 \ \text{ó} \ B=10$$

An analysis shows that

$$\sqrt{57-40\sqrt{2}}<\sqrt{57+40\sqrt{2}}$$

that is to say

$$\sqrt{57-40\sqrt{2}}-\sqrt{57+40\sqrt{2}}<0,$$

Then, $B=-10$.

Therefore, $(A+B)^2=(3-10)^2=49$

Exercise 5 is an example of mathematical situations whose solution path does not fully correspond to the contents studied by the student, where the student must show more creative ingenuity than deep mathematical knowledge.

A possible solution is WATER = 5175, considering A=5, T=3 , U=7, 0=0 and G=1, i.e. 1035+1035+1035+1035+1035 = 5175.

TEST 2: Inter-school (municipal) or inter-municipal (provincial) phase

1. ^What is the sum of the four prime divisors of the number $K=2^{16}-1$?
2. Let $a^3-b^3=24$ y $a-b=2$. Determine the numerical value of $(a+b)^2$.
3. The side, the height and the area of an equilateral triangle inscribed in a circle form, in this order, a geometric progression. Calculate the radius of the circle.
4. Proof that $\forall$ a, b, c $\in \mathbb{R}$, holds that

   $a^2+b^2+c^2\geq ab+ac+bc$
5. (OPM, 2016) The natural numbers are coloured green or blue so that:

- The sum of a green and a blue is blue;
- The product of green and blue is green.

How many ways are there to colour the natural numbers with these rules, so that 462 is blue and 2016 is green?

In this second test, the first three exercises can be considered as POEDs, and the last two as POEIDs. Let us look at the procedures for solving each exercise:

Resolution of exercise 1:

$K = 2^{16} - 1$
$K = (2^8 - 1)(2^8 + 1)$
$K = (2^4 - 1)(2^4 + 1)(2^8 + 1)$
$K = (2^2 - 1)(2^2 + 1)(2^4 + 1)(2^8 + 1)$
$K = (2 - 1)(2 + 1)(2^2 + 1)(2^4 + 1)(2^8 + 1)$
$K = 1 \times 3 \times 5 \times 17 \times 257$

The sum S_d of the prime divisors of K is

$$S_d = 3 + 5 + 17 + 257 = 282$$

Resolution of exercise 2:

$$a^3 - b^3 = (a - b)(a^2 + ab + b^2)$$

$$24 = 2\,(a^2 + ab + b^2)$$

$$a^2 + ab + b^2 = 12 \qquad (1)$$

As $a - b = 2 \quad \Rightarrow a = 2 + b,$ substituting this expression in (1),
We obtain

$$(2 + b)^2 + (2 + b)b + b^2 = 12$$

$$4 + 4b + b^2 + 2b + b^2 + b^2 = 12$$

$$3b^2 + 6b - 8 = 0$$

From the quadratic equation, we get

$$b_1 = -1 + \frac{\sqrt{17}}{3} \quad ó \quad b_2 = -1 - \frac{\sqrt{17}}{3}$$

Hence $a_1 = 2 + b_1 = 1 + \frac{\sqrt{17}}{3}$ e $a_2 = 2 + b_2 = 1 - \frac{\sqrt{17}}{3}$, i.e. the *S-solution.*
of the quadratic equation are two cases:

$$S_1 = \left\{\left(1 + \frac{\sqrt{17}}{3}, -1 + \frac{\sqrt{17}}{3}\right)\right\}$$

$$S_2 = \left\{\left(1 - \frac{\sqrt{17}}{3}, -1 - \frac{\sqrt{17}}{3}\right)\right\}$$

Developing the binomial

$$(a + b)^2 = a^2 + 2ab + b^2 = a^2 + ab + b^2 + ab = 12 + ab$$

For case $S1$, we have:

$$(a_1 + b_1)^2 = 12 + a_1 b_1$$

$$12 + \left[-\left(1 + \frac{\sqrt{17}}{3}\right)\left(1 - \frac{\sqrt{17}}{3}\right)\right]$$

$$12 + \left[-\left(1 + \frac{\sqrt{17}}{3}\right)\left(1 - \frac{\sqrt{17}}{3}\right)\right]$$

$$12 - \left(1 - \frac{17}{9}\right) = \frac{116}{9}$$

For the case $S2$, we have:

$$(a_2+b_2)^2 = 12 + a_2 b_2$$

$$12+\left[-\left(1-\frac{\sqrt{17}}{3}\right)\left(1+\frac{\sqrt{17}}{3}\right)\right]$$

$$12-\left(1-\frac{17}{9}\right)=\frac{116}{9}$$

Solution: $(a+b)^2=\frac{116}{9}$

Resolution of exercise 3:

Sea $P=\{lado, altura, area\}=\left\{l,\frac{l\sqrt{3}}{2},\frac{l^2\sqrt{3}}{4}\right\}$

Since P is a geometric progression, then

$$\frac{\frac{l\sqrt{3}}{2}}{l}=q \quad (1)$$

Y

$$\frac{\frac{l^2\sqrt{3}}{4}}{\frac{l\sqrt{3}}{2}}=q \quad (2)$$

From expression (1) it follows that

$$q=\frac{\sqrt{3}}{2}$$

From expression (2) it follows that

$$q=\frac{l}{2}$$

Later, $l=\sqrt{3}$

The radius of a circle inscribed in a triangle is $r=\frac{2}{3}\cdot height$

Substituting the height in this expression, we obtain:

$$r=\frac{2}{3}\cdot\frac{l\sqrt{3}}{2} \Rightarrow r=\frac{2}{3}\cdot\frac{\sqrt{3}\cdot\sqrt{3}}{2} \Rightarrow r=1$$

Solution: The radius of the circle measures 1.

Resolution of exercise 4:

Demonstration: Let $a,b,c\in\mathbb{R}$.

$$\left.\begin{array}{l}\forall\, a,b,\in\mathbb{R},\ (a-b)^2\geq 0\\ \forall\, a,c\in\mathbb{R},\ (a-c)^2\geq 0\\ \forall\, b,c\in\mathbb{R},\ (b-c)^2\geq 0\end{array}\right\}\Rightarrow\quad \begin{array}{l}a^2+b^2-2ab\geq 0\\ a^2+c^2-2ac\geq 0\\ b^2+c^2-2bc\geq 0\end{array}\quad (1)$$

Adding member by member in the system (1), it remains:

$$2(a^2+b^2+c^2)-2(ab+ac+bc)\geq 0$$

Later,

$a^2 + b^2 + c^2 \geq ab + ac + bc$ □

<u>Resolution of exercise 5</u>:

First we note that the number 1 must be painted blue. In fact, if 1 is green, given a blue number A, as $1 \times A = A$, we conclude, according to the second rule, that A is green, which is false. Therefore, 1 must be blue.

Let V_1 and V_2 be two numbers painted green, then, using the first rule and the fact that 1 is blue, $V_1 + 1$ is a blue number. Thus, by the second rule, $V_2 x (V1 + 1) = V1 x V_2 + V_2$ is green, from which it follows that $V1 x V_2$ is also green, so as not to contradict the first rule. This argument, together with the second rule, proves that the product of a green number and any other natural number is always green. In other words, all multiplies of a number painted green are numbers painted green.

Let v be the smallest of all natural numbers coloured green. We have already seen that $v \neq 1$. We can easily argue that $v \neq 2$. Because if 2 were green, then $462 = 2 \times 231$ would also have to be green, which is not the case. Similarly, $v \neq 3$.

We know that $v, 2v, 3v, \dots, kv, \dots$, are all numbers painted green. Let us now show that these are the only natural numbers painted green. By definition of v, we know that all the natural numbers less than v

$1,2, \dots, v-2, v-1$

1 are blue. Adding v to each of these natural numbers gives the numbers

$v+1, v+2, \dots, 2v-2, 2v-1$

are also all blue, according to the first rule. Thus we have shown that all the natural numbers strictly contained between v and $2v$ are blue. Adding v again, we show that all the natural numbers strictly contained between $2v$ and $3v$ are blue. So on, it is shown that the only natural numbers painted in green are the multiplies of v.

To paint the natural numbers according to the statement, then we have to choose a natural number v that divides 2016 but does not divide 462. The number $2016 = 2^5 \times 3^2 \times 7$ has 36 divisors, all of the form:

$2^a \times 3^b \times 7^c$, con $a \in \{0,1, \dots ,5\}$, $b \in \{0,1,2,3\}$ y $c \in \{0,1\}$.

Of these 36 divisors, there are 8 that are also divisors of 462 = 2 x 3 x 7 x 11.

$2^a \times 3^b \times 7^c$, con $a \in \{0,1\}$, $b \in \{0,1\}$ y $c \in \{0,1\}$.

Solution: There are 36 - 8 = 28 natural numbers v in the given conditions and each of these natural numbers v defines a way of representing the natural numbers that respects the rules of the statement.

Proofs 1 and 2 are only examples of how to combine in their structure the two types of *problems eminently deductive* and *eminently inductive-deductive*. It must be taken into account that for the construction of a Mathematical Olympiad test, for both types of problems, the problems must be interesting, original, which cannot be very similar to those that the contestant could or could not solve in his preparation, in such a way that he does not have a clear idea of their solution a priori.

4. Measurement indicators for the identification of gifted students in mathematics competitions.

There are several models that have indicators designed for the identification of gifted learners. According to (Renzulli, 2015) cited by (Garrta, Torres & Torres, 2016), scientific findings in recent decades support the idea of a broader system for identifying students with high intellectual ability. Most researchers and practitioners agree that a single intelligence or achievement test score is no longer sufficient. The first and most important decision that should be taken in relation to the implementation of an identification model should be what conception or definition of high intellectual ability is to be adopted in a particular school and what attention it is intended to provide.

For his part, (Williams, 1981) cited by (Ramos Palacios, 2006) states that "talented individuals are defined in the context in which they act and that talent is relative and depends on geographical, temporal and cultural variables that change according to the time".

With regard to the selection of young people with mathematical skills from the Olympiads held in Angola, it is worth noting that article 20 e) of the regulations of the Olympiad Mathematics Olympiad stipulates that the winners are those contestants who achieve the highest number of correct answers within the stipulated time and whose score is equal to or greater than 65% of the total, as can be seen in point 3 of article 7 of the said regulations.

This criterion is typical of a competition, but from our point of view, it is not sufficient to identify young talents in mathematics, because, as we know, winner is not synonymous with talent. On the other hand, this criterion may allow the organisers of the Olympiad at each of its stages to deviate in relation to the identification of current talents and to enhance talents in mathematics.

Practice indicates that the lack of evaluation criteria can lead teachers who act as juries for the different stages of the Olympiad to qualify the contestants only by measuring the correct answers, with the risk of forgetting students with high abilities who did not reach the podium, as well as not detecting the main difficulties of the students. In this regard (Castillo & Cabrerizo, 2009) "measuring is a necessary condition for evaluating, but not sufficient", so evaluating is more complex as it involves an interpretation of the measures referenced to a series of criteria related to the objectives (Moreno, Penalosa & Cueto, 2016).

Hence the need to establish indicators and criteria for the mathematics olympiads in the schools of the province of Huambo, depending on the tests to be applied, which will make it possible to identify talented pupils, whether winners or not, in order to give them a differentiated attention for their qualitative progress in mathematics. The indicators are thus proposed:

- In the intra-group and inter-group Olympiads: the pupil who is able to solve at least 75% of the eminently deductive Kempic problems and 50% of the eminently inductive-deductive Kempic problems or who is able to solve at least 50% of the eminently deductive Kempic problems and 100% of the eminently inductive-deductive Kempic problems will be considered talented.
- In inter-school Olympiads within the same municipality and inter-municipal (provincial) Olympiads: a pupil who is able to solve at least 50% of the eminently

deductive oKmpic problems and 75% of the eminently inductive-deductive oKmpic problems or, a pupil who is able to solve at least 50% of the eminently deductive oKmpic problems and 100% of the eminently inductive-deductive oKmpic problems will be considered talented.

The indicators and criteria of evaluation of the tests of the Olympiad "Olympiad of Mathematics", constitute a valuable instrument for the teachers to be able to have evaluative judgments, with a globalizing appreciation for the selection of the talented pupils, to diagnose the participants and to stimulate their results from the distribution of the points.

As for the assessment criteria, understanding that solving mathematical problems is a heuristic and not purely algorithmic act that involves various phases, (Callejo, 1996) cited by (Pons, 2017) concludes that the assessment of student learning in problem solving should not be limited to the assessment of the results or the point of arrival of the students, as it should also consider the processes of searching for the solution to the problem.

Considering that the olympiad tests, which combine in their structure eminently deductive and eminently inductive-deductive oKmpic problems, are instruments that emanate from the evaluation as they make it possible to obtain data on the mathematical knowledge and skills that the pupils are acquiring in their academic training; Also taking into account that it is a process of globalising assessment, that is, a holistic assessment based on a single instrument, in order to assess the resolution of each Olympic problem, the author adopts an assessment matrix for the Angolan context of measurement, which is shown in the following table:

DESCRIPTION		**EVALUATION**	
		Weight	**Score**
Solve the problem	Error-free, demonstrating mastery of the skills necessary to achieve the result.	100%	4
	In the essential aspects, making elementary procedural and/or calculation errors,	80%	3,2
Partially solves the problem	It includes in the resolution process most of the requirements of the original task.	60%	2,4
	It includes in the solution process essential requirements of the original task in correspondence with the errors of understanding that I manifest in the solution of the problem.	40%	1,6
It does not solve the problem,	But in the attempt to solve the problem, it evidences mastery of the content and skills necessary to reach the correct solution.	20%	0,8
	None of the above situations apply	0	0

5. The reports as a result of the Mathematics competitions

The production of reports allows to control the process of the realisation of the Mathematics Olympiad. It is from the reports that some of the difficulties of the PEA of Mathematics in the schools of Huambo will be known and, with that to conceive actions that lead to the improvement of the DCM.

Thus, the juries of each stage of the Olympiad must produce a report detailing the participation of the contestants by gender, age and referencing the topics in which the students in general have the greatest difficulties, and attach to the report the test

applied with the respective keys.
Such reports should be sent to the immediate Mathematics co-ordinators of each stage of the Olympiad, i.e. the report of the intra-school and inter-school Olympiads will be sent to the school Mathematics discipline co-ordinator and the reports of the inter-school (municipal) Olympiads will be sent to the provincial Mathematics discipline co-ordinator.
The reports of each stage can be used as a basis for the creation of a magazine, which in turn can serve as a support for the self-preparation of students wishing to compete in the mathematical olympiad. In addition, the reports encourage the juries to ensure scientific rigour in the preparation of the Olympiad tests.
Analysing the importance of the report as a result of the Mathematics Olympiad, we have:
S *Reflection of the problem solving process:* The reports allow to document and show the process followed by the participants to solve the problems posed in the Mathematics Olympiad. This provides a detailed and transparent view of how students apply different strategies, reasoning and skills.
mathematics in their search for solutions. These reports can be studied by researchers and mathematics educators to better understand the thinking and approach to problems of the student competitors.
S *Post-competition analysis and reflection:* The reports are a valuable source of information for post-competition analysis and reflection. Mathematics Olympiad officials can use the reports to evaluate and improve the quality of the problems posed, to detect patterns of thinking or strategies used by students and, in general, to analyse the results and trends observed in the solutions.
S *Individualised feedback:* The debriefings provide participants with individualised feedback on their performance in the Mathematics Olympiad. These feedbacks are valuable, as students can understand where their solutions were correct, where they may have made mistakes and how they can improve in future competitions or mathematical challenges.
S *Learning tool:* The relays serve this purpose. As contestants analyse each other's solutions, they learn different approaches, strategies and methods of problem solving, which encourages metacognitive reflection, the broadening of mathematical knowledge and the development of critical and creative skills.
S *Promotion of communication and collaboration:* Through the relatorias, participants can communicate, share and discuss their mathematical ideas. This fosters collaboration and knowledge sharing between teachers and contestants, creating an enriching and motivating mathematical community. Effective communication around mathematical solutions facilitates joint knowledge building and academic growth.

6. Differentiated educational attention and the preparation of gifted pupils in mathematics competitions.

In Angola, mathematics competitions are part of the extra-curricular activities, as they form part of the school calendar of the General Education Subsystem, which is published annually by means of an executive decree of the Ministry of Education. This means that the preparation activities for the contestants, especially for gifted pupils, are seen in the context of differentiated educational attention.

In the field of mathematics competitions, differentiated educational attention implies adapting teaching content, methods and activities to the needs and abilities of gifted pupils.

From the approach of Lev Vygotsky's psychology, the differentiated educational attention and the preparation of gifted pupils in Mathematics competitions are based on the following theoretical and methodological foundations:

S *Zone of Proximal Development (ZDP)*: Vygotsky proposed the idea of the Zone of Proximal Development, which refers to the space between what a student can do independently and what he/she can do with the assistance of an adult or a more competent peer. In the case of differentiated educational attention, the aim is to identify and provide gifted students with challenges and tasks that are in their ZDP in order to foster their mathematical growth and development.

S *Scaffolding*: In line with the ZDP, the concept of scaffolding involves providing support and guidance to students as they tackle more advanced tasks. Gifted students receive structured support and assistance from teachers or expert mentors in preparation for the Math Olympiad. As students acquire additional mathematical skills and knowledge, the scaffolding is gradually modified and decreased, allowing them to develop autonomy.

J *Collaborative learning*: Vygotsky emphasised social learning and the importance of peer interaction in the learning process. In differentiated educational attention, gifted students can benefit from collaboration and teamwork with other students of similar abilities. Through discussion and joint problem solving, gifted students have the opportunity to share and build mathematical knowledge collaboratively.

J *Mediation and cultural tools*: Vygotsky emphasised the role of mediation and cultural tools in cognitive development. In the case of the preparation of gifted students, different resources and materials are used to facilitate their learning and development in mathematics. This may include access to advanced books and materials, the use of technology for learning, and the presentation of challenging and relevant mathematical problems.

J *Use of language and dialogue*: Vygotsky stressed the importance of language and dialogue in learning. In differentiated instruction, dialogue and active communication are encouraged among students and with teachers or mentors, so that they can express their thoughts, share ideas and receive feedback. Mathematical language and conceptual discussion facilitate the internalisation of knowledge and the development of more sophisticated mathematical skills.

As it has been expressed, the preparation of contestants cannot be seen as an isolated element within the pedagogical process in the educational environment, it is part of it and maintains the social character of education. This means that this preparation process contributes to the integral development of all the contestants, in a dialectic way, where the social and the individual are linked, where space is created to attend to those with high capacities, but in accordance with the ideal of the man to be formed and who integrates himself into his social context in a creative way.

Mathematics, as Jungk (1979) states, must develop qualities such as: dedication, constancy, steadfastness, perseverance, the will to overcome difficulties, criticism and self-criticism, and collective behaviour. In the preparation of the contestants, in particular, the incidence of motivation plays a fundamental role in this process, since the student, when faced with exercises and problems that require a greater effort, many become frustrated at not being able to fulfil the task as they usually do in the curricular exercises.

In this regard Bravo (s.a), according to (Perez, 2014), states that motivation is an important didactic resource, which acts in favour of learning, as it increases understanding and enhances the study process, confronts the student and provides effectiveness to teaching.

Generally, in the first moments of the contestants' preparation, this is carried out under the direction or tutelage of the teacher or trainer who stimulates and motivates them, but other contestants of higher grades also intervene and provide help through their experience, until in a short time they reach the independence necessary for their self-preparation. This also reveals the essence where learning requires the transition from dependence to independence and self-regulation of the subject, transforming himself and influencing his environment, since, being in higher stages of learning, he also helps others.

In the preparation group, the mathematics competitors exchange their experiences and the knowledge of the members of the group is enriched to the same extent that important convictions and qualities of the personality of the students are developed under the guidance of the trainers, who are responsible for modelling the preparation of each mathematics competitor according to his or her particularities on the basis of the diagnosis they have of him or her.

The trainer, for his part, in order to model the preparation of each contestant, has to achieve the unity of the didactic functions, interweaving the old with the new, according to Klingberg (1978). This process of preparation itself is a process of teaching-learning of Mathematics, a didactic process, which is generally carried out in a space of time outside or inside the curriculum, during school hours or not, where the contestants receive this preparation.

Internationally, three main ways of developing the preparation of candidates are identified: segregation, acceleration and curricular enrichment.

According to Lorenzo and M. Martmez (2009), segregation can be manifested in a partial or total way. Partial segregation consists of separating contestants from the regular group at certain times, or days of the week, to interact with peers in special activities. This form is one of the most widespread and feasible to apply in the particular case of Angola, since the contestant receives his or her preparation in an extracurricular schedule.

In total segregation, special schools or classrooms are created. In this form there is usually a trainer or a team of trainers to carry out the process. In Angola there are no such centres and the conditions for their development are not yet in place.

The same authors define acceleration as the organisation of a programme, a subject or a grade so that contestants can complete a programme, a subject or a grade in less time than the time stipulated for average students. This involves adapting the curriculum and making special plans, since the regular teaching time is reduced in order to achieve the general objectives of the programme, subject or grade, and to receive additional preparation in the contents of the mathematics competition. In this

case the student may or may not remain in his or her regular group. The same Mathematics teacher who teaches the subject may or may not be the coach, depending on the extent to which the acceleration occurs.
This model requires a high level of preparation on the part of the teacher or trainer. In Angolan conditions, where most teachers are not adequately trained, it is almost impossible to accelerate the contestant's performance.
Enrichment, on the other hand, consists of the student remaining in his or her regular group and being offered a variety of activities in addition to the regular programme. The latter is the most widely used and accepted way to stimulate their development, as it does not involve any psychological risk for the student and is a more flexible option. Like the accelerated model, it requires a high level of teacher preparation.
One of the most widely accepted conceptions of curriculum enrichment is, according to Apraiz (2002), the Renzulli model, with its three levels. Level I includes proposing to students new and interesting topics, ideas and fields of knowledge, additional to the regular curriculum. Level II involves proposing activities that develop critical and creative thinking to solve problems; learning skills, such as classifying, analysing data or drawing conclusions; skills in using bibliographical sources; oral communication skills. Level III consists of developing investigations of real problems individually or in small groups. The aim is to achieve the application of knowledge, creativity and motivation for a free topic and to acquire higher level knowledge and methods within a specific field.
In the forms of preparation where curricular enrichment predominates, M. D^az (2009) also proposes partial segregation in the preparation of Mathematics contestants. On the other hand, M. Martmez (2009), evaluates a vitally important element, the possibility of using the three forms of preparation of talented students in a concatenated manner. In their respective articles, D. Castellanos and C. Vera (2009);
Lorenzo and M. Martmez (2009); M. Martmez (2009); and M. D^az (2009), state that knowledge quizzes are effective means to prepare them. Only M. D^az (2009) refers to the need to periodically evaluate their knowledge in order to consolidate the idea of competition among students in the group. This author sustains the importance of preparation in the independent elaboration and solution of problems by the contestants.
Under Angola's conditions, the way in which the preparation of the contestants is carried out is partial segregation, since the contestants will be separated from the regular group at certain times and days of the week, to interact with the coach and other peers. This can be done from the classroom level up to any of the competitive levels as a selective process, although at the classroom level, curricular enrichment is evident mainly at levels I and II.
The educational agents in charge of the preparation of the pupils at each stage of the competition "Olympiad of Mathematics" are recommended below.
- In the intra- and inter-group Olympiad, the differentiated attention and preparation of pupils should be carried out by the teachers who teach in the respective groups, i.e. in the framework of curricular and extra-curricular activities. In this respect (Palacios, 2006) points out that:
"The importance of identification in the educational field is very significant, as it

allows us to detect in time the specific talents or potentials of students and thus to be able to attend to their special educational needs. In this process the teacher is a valuable source of information".

- In the inter-school (municipal) olympiad, it will be up to the mathematics co-ordination of the municipality to indicate the teachers who will be in charge of the preparation of the winners of this phase with a view to a good performance in the next phase. In the absence of a teacher versed in Olympiad training, this function should be carried out by the teachers whose pupils have won this phase, as they have the advantage of knowing many of the pupils' potentialities and the aspects on which they should focus.
- In the provincial olympiad, it is up to the provincial mathematics co-ordinating office to appoint teachers to prepare the winners of this phase for the next phase. On the other hand, in the absence of a teacher with expertise in Olympiad training, this function should be carried out by the teachers whose students were the winners of this phase.

It is important to underline that the identified talents, even if they are not winners at a certain stage of the competition, must continue and deserve a differentiated attention in the academic environment by their respective teachers, as these students continue to be talents within their context (Villarraga, Martmez and Benavides, 2004) reveal that:

"Talent has an evolutionary character in the sense that not only the actual talent of an individual is relevant, but also the potential talent is fundamental, because from this it is possible to make interventions to foster and develop talent".

In cases where the mathematics co-ordinating office indicates teachers to prepare students for the next stage of the Olympiad, the selection should be made by experienced teachers. In this regard (Feldhusen, 1997) cited by (Conejeros Solar, Gomez Arizaga and Osorio, 2011), argued that the teacher for high abilities should have some of the same characteristics and skills of their gifted students and in this line indicated that teachers working with gifted students should also have special skills and knowledge regarding the particular characteristics of these children that facilitate their personal, social and academic development.

Teacher trainers of gifted pupils should have the following characteristics:

S *Proficiency in mathematics:* Teachers who train students for Mathematics competitions must have a broad command of mathematical concepts. They should have a thorough knowledge of the theory and be familiar with the different approaches and techniques used in quiz competitions.

S *Competition experience:* If possible, it would be beneficial for teachers to have previous experience of mathematics competitions, particularly in secondary education or in their professional training, without making demands on the quality of their results. This undoubtedly enables them to understand the peculiarities of the problems and the strategies used to solve them. Experience from previous competitions helps them to provide guidance and practical advice to students.

S *Ability to teach strategies:* Teachers must be able to teach students the problem-solving strategies and techniques expected for mathematics competitions. This involves teaching them how to analyse and decompose difficult problems into smaller steps, identify patterns and apply mathematical concepts creatively. In short, teaching them to use heuristic rules, strategies and principles.

S *Motivation and enthusiasm:* Teachers who train students for mathematics competitions must be motivating and enthusiastic. They must be able to inspire students, generate curiosity towards mathematical challenges and foster a positive learning environment.

S *Communication skills:* It is important that teachers are able to communicate clearly and effectively with students. They must be able to explain mathematical concepts in an understandable way and establish a good rapport with students to stimulate their participation and confidence.

J Flexibility and adaptability: Every learner is different, so teachers must be flexible and adapt their teaching approach to the individual needs of each learner. They must be able to identify the strengths and weaknesses of each student and personalise training accordingly.

Regarding the aspects that should be considered in the curricula for gifted students, Feldhusen and Wyman cited by (Pacheco Urbina, 2001) present their criteria of basic needs, which will be taken into account when dealing with the mathematical talents identified from the olympiads in Angolan Secondary Education:

J Maximum achievement of basic skills and concepts.

J Learning activities at an appropriate level and pace.

J Experiences in creative thinking and problem solving.

J Development of convergent skills, especially logical deduction and problem solving.

J Stimulation of imagination and spatial skills.

J Development of self-awareness and acceptance of one's own abilities, interests and needs.

J Encouragement to pursue high-level goals and aspirations.

J Development of independence, self-direction and discipline in learning.

J Experiences of interacting with other intellectually, artistically and effectively highly skilled, creative and talented students.

J Extensive sources of information on various topics.

J Access and stimulation for reading.

In the case of Olympic medals (gold, silver or bronze), it is suggested that these be awarded only to talented winners according to whether they come first, second or third. This measure may encourage the winners without a medal to further prepare for the next stage of the Olympiad.

The other prizes should be such that they contribute to the intellectual development of the winners.

On the other hand, at each stage of the Olympiad the coaches must provide the winners with training with a greater emphasis on Olympic-type problems with a view to performing well at the next stage where the weight for this type of exercise is greater.

CHAPTER 5

Challenges and changes needed to improve performance in mathematics contests

The author's teaching experience in the preparation of the coordinators of the discipline of Mathematics for the competitions and 12 years of teaching work in the province Huambo, has allowed him to consider the following challenges to improve the performance of the province in the Mathematics Olympiads:

> The constant improvement of the teachers *in the didactic aspects that serve as a basis for the materialisation of the objectives of the national competition "Olympiads of Mathematics".* Developing in teachers:

- Skills for the conception of efficient teaching methods in the conduct of the training of the contesting students, which is a process of teaching-learning of Mathematics.
- Skills for assessing student competitors (the structure of the tests, the types of problems and assessment criteria).

> Continuous professional updating of teachers, since the olympiads are an element that favours the improvement of educational systems in that they involve a permanent updating of teachers' knowledge, a search for new problems and methods of adapting new and more attractive content to existing curricula.

> To allow teachers with high skills in teaching mathematics to contribute their knowledge in the coordination of this school discipline, without exclusion of any kind. It is necessary to make proper use of human resources by making a greater effort in educational tasks, and to use the scientific and technological advances available to us, as well as for the authorities to reconsider their partisan vision in the positions of head of the teaching activity.

> The projection and development from the school of a differentiated training system for students who wish to participate in the secondary school mathematics competition in the province of Huambo.

> To develop a movement of promotion and motivation for mathematics as a cultural manifestation, forming "Friends of Mathematics" groups interested in deepening their mathematical knowledge and their participation in competitions, for which support should be sought in different ways, involving managers, teachers, students, parents, press, social networks and other institutions, this support should be of accompaniment, moral support and material stimuli.

> Another fundamental challenge is the government's low support to the education sector. It is of utmost importance that Angola's allocations for education should be increased to the 20% stipulated in its international commitments, in order for the country to achieve the 4th Sustainable Development Goal, that of "Quality Education".

Bibliography

Andre, M. A. C. D, (2016): *La superacion profesional del docente en tratamiento metodologico del contenido matematico.* Thesis in Opcion al Grado Cientifico de Doctor en Ciencias Pedagogicas en Universidad Central "Marta Abreu" de las Villas, Santa Clara - Cuba.

Angola. Instituto Nacional de Investiga^ao e Desenvolvimento da Educagao, (2013). *Curriculo do 2° Ciclo do Ensino Secundario Geral.*

Angola. Instituto Nacional de Investiga^ao e Desenvolvimento da Educagao, (2013). *Cumculo do 1° Ciclo do Ensino Secundario.*

Angola. Lei n° 13/01 de 31 de Dezembro de 2001. *Lei de Bases do Sistema de Educagao.*

Angola. Lei n° 17/16, de 7 de Outubro de 2016. *Lei de Bases do Sistema de Educagao e Ensino de Angola.*

Angola. Lei n° 32/20, de 12 de Agosto de 2020. Lei que altera a Lei n° 17/16 de 7 de Outubro - *Lei de Bases do Sistema de Educagao e Ensino.*

Angola. Ministerio da Economia e Planeamento (2018). *Plano de Nacional de Desenvolvimento 2018 - 2022.*

Angola. Ministerio da Educagao (2018). *Estatuto da Carreira dos Agentes de Educagao.* Decreto Presidencial n°. 160/18 de 3 de Julho. Imprensa Nacional.

Angola. Ministerio da Educagao (2018). *Regulamento da Olimpiadas de Matematica.* Decreto Executivo n°. 03/18 de 15 de Maio.

Angola. Ministerio da Educagao (2021). *Regime juridico do exercicio do cargo de direcgao e chefia em instituigoes de Educagao Pre-escolar, Ensino Primario e Secundario.* Decreto Presidencial n°.93/21 de 16 de Abril.

Apraiz de Elorza, J. (2002). *La educacion del alumnado con altas capacidades.*

Azcarate, P. (2006). *Alternative proposals for evaluation in the mathematics classroom.* In J.M. Chamoso (Ed.), Enfoques actuales en la didactica de las Matematicas. Madrid: MEC, Coleccion Aulas de Verano, 2006, p. 187-219.

Ballester Pedroso, S et al. (1992). *Metodologia de la Ensenanza de la Matematica Tomo I.* Cuba. Editorial Pueblo y Educacion

Ballester Pedroso, S. G. (2018). *Didactica de la Matematica.* Havana: Felix Varela.

Caceres, S. G. C.; Reyes, L. I. M. & Hofmann, G. O (2016). *Formative assessment in Mathematics. Strategies and instruments.*

Caimbo, N. G. (2013). *Scientific conception for the pedagogical and didactic management of the teaching-educational process at the University of Lueji A 'nkonde of the Republic of Angola* (Doctoral thesis). University of Pinar del Rfo, Cuba.

Cameira, A. F. P., Rodriguez, J. M. R., Garda, G. G. (2018). *Desenvolvimento profissional e formagao continua de professores: uma analise do contexto da educagao em Angola.* INNOEDUCA. International Journal of Jechnology and Educational Innovation Vol. 4. no. 1. June 2018 pp. 71-78.

Carazo A, Norberto J.; Carazo A., Alfredo B. (2016): *Teacher training for mathematics competitions in secondary education.* Atenas, vol. 3, num. 35, 2016 Universidad de Matanzas Camilo Cienfuegos, Cuba.

Cardoso, E. (2012). *Problemas e Desafios na Formagao Inicial de Professores em Angola. Um estudo nos ISCED da Regiao Academica II* (Doctoral Thesis). Universidade de Minho, Minho, Portugal.

Cassinda, L. (2014). *La superacion didactica de los profesores para la formacion de habilidades cientificas investigativas en los estudiantes del segundo ciclo de la ensenanza secundaria en Huambo, Angola (*Doctoral thesis), UCPEJV, Havana.

Castellanos Simons, D., and Vera Salazar, C. (2009). *Educational intervention for the* *development of talent at school. In D. Castellanos Simons (Comp.). Talento: concepciones y estrategias para su desarrollo en el contexto escolar (*pp. 83102). Havana: Pueblo y Educacion.

Castillo, S. & Cabrerizo, J. (2009). *Evaluacion educativa de aprendizajes y competencias.* Spain.

Castro Escarra, Olga (1997). *Fundamentos Teoricos y metodologicos del Sistema de Superacion del Personal Docente del Ministerio de Educacion,* (Tesis de Maestria en Educacion Avanzada), ISPJV, La Habana, Cuba.

Conejeros-Solar, M. L., Gomez-Arizaga, M. P., Osorio, E. D. (2011). *Teaching profile for students with high abilities.* Magis. International Journal of Research in Education, vol. 5, num. 11, January-June, 2013, pp. 393 - 411. Pontificia Universidad Javeriana Bogota, Colombia.

Crespo Hurtado, E. T. (2007): *Modelo didactico sustentado en la heuristics para el proceso de ensenanza-aprendizaje de la Matematica asistida por computador.* [Doctoral thesis, Universidad Pedagogica "Felix Varela"].

Cruz, M. (2006). *La ensenanza de la matematica a traves de la resolution de problemas (Vol. 1).* Instituto Superior Pedagogico "Jose de la Luz y Caballero", Havana, Cuba.

Chiumbo Paiva, J., Crespo Hurtado, E., Lopez Fernandez, R., Crespo Borges, T. (2020). *Mathematics contests in Huambo province: a pedagogical model for their development. Revista Conrado, 16*(S 1), pp. 249-255.

Chiumbo Paiva, J., Lopez Fernandez, R., & Crespo Hurtado, E. (2023). *Proposta de uma estrutura geral e criterios de avaliaqao para as provas dos Concursos de Matematica na provincia do Huambo.* Revista Conrado, 19(S1), 129-136.

Da Costa, R. M., & Kunjiquisse, J. A. (2012). *Pedagogical improvement system to enhance the professional pedagogical performance of teachers of Higher Education Centres in Huambo" - Angola.* University Congress, Vol. I, No. 3, ISSN: 2306-918X., 395-415.

De Almeida, J. F. (2015). *Strategy of pedagogical professional improvement in mathematical problem solving for the Middle Industrial Institutes of Angola (*Doctoral thesis), ICCP, Havana.

Delgado, N. (2009). *The preparation of teachers to work with students attending the national chemistry olympiads.* Congreso Internacional Pedagog^a. Havana.

D^az Gonzalez, M. (2002, May). *The preparation of talented students in Mathematics in Cuba. Meeting with students and teachers of the National Training Centre.* Havana, Cuba.

D^az Gonzalez, M. (2009). *Attention to gifted students in the subject of Mathematics. Some notes. In D. Castellanos Simons (Comp.) Talento: concepciones y estrategias para su desarrollo en el contexto escolar (*pp. 1972-10). Havana: Pueblo y Educacion.

Principals Making School, Michailuk, M. C. & Nicodemo, M. (2015). *La evaluación en el área de matematica. Claves y Criterios. Nivel Secundario.* OEI, Buenos Aires.

Drago, C. (2017). *Manual de Apoyo Docente. Evaluacion para el Aprendizaje.* Chile.

Escobar, G. (2000). *Acciones de superacion para capacitar a profesores de matematica del nivel medio, en la ejecución del entrenamiento de los concursos estudiantiles. Unpublished master's thesis.* Santa Clara: University of Pedagogical Sciences "Felix Varela".

Espinosa, F. O. (2005). *Estrategia de Capacitacion para profesores de preuniversitarios, en funcion del desarrollo de talentos matematicos.* [Master's thesis, Universidad de Matanzas Camilo Cienfuegos].

Fernandes, A. S. (2021). *Resolugao de problemas olimpicos envolvendo analise combinatoria e probabilidade atraves da Metodologia de Polya.*

Gangula, E. W., Faustino, A. (2018). *Dilema da formagao matematica em Angola: falta de iniciativas proprias ou de compromisso com a qualidade de ensino?* Journal Actualidades Investigativas em Educação. Volume 18, n° 3, pp. 1 - 22.

Garcia, F, F. (1997). *Evaluation of competences in Elementary Algebra through verbal problems.* [Doctoral Thesis, University of Granada, Spain].
Gonzalez, Z. (2007). *La preparacion del maestro de la escuela primaria para la realizacion efectiva del diagnostico integral del escolar. (*Doctoral thesis) Universidad de Ciencias Pedagogicas "Felix Varela". Villa Clara, Cuba.
Haydt, R. C. C. (2011). *Curso de Didactica Geral.* Sao Paulo. Editora Atica.
Ibanez, C. M. A. & Boada, M. J. A. (2016). *Evaluacion En Matematicas. Una Propuesta Basada en Competencias para el Colegio de Bachillerato Patria.* Colombia.
Jimenez, D., Gonzalez, J., & Tornel, M. (2018). *Formacion del profesorado universitario en metodologias y su incidencia en el aula. Pedagogical Studies XLIV.*
Juliao, A. L. (2020). *Formagao de professores, ensino primario e qualidade educativa em Angola: vazios e pontes na relagao.* Revista Internacional de Formagao de Professores (RIFP), Itapetininga, v. 5, e 020002, p. 1-20.
Jungk, Werner (1979). *Conferencias sobre metodologia de la ensenanza de la Matematica I.* Pueblo y Educacion, Havana.
Junior, M. P. O., Pinheiro, H. M., Barreto, W. D. L. (2022). *Um estudo de caso sobre a aplicagao das técnicas de soluugao de problemas de Olimpiadas de Matematica para a melhoria do ensino da disciplina.* Research, Society and Development, 11(6), e53611629295, 2022 (CC BY 4.0).
Klingberg, L. (1978). *Introduccion a la Didactica General.* Havana: Editorial Pueblo y Educacion.
Liberato, E. (2014). *Avangos e retrocessos da Educagao em Angola.* Revista Brasileira de Educa?ao, 19(59) , Out. - Dez 2014, pp. 1003 - 1031.
Lorenzo Gartia, R. and Martinez Llantada, M. (2009). *Polemics around the development of talent. In D. Castellanos Simons (Comp.) Talento: concepciones y estrategias para su desarrollo en el contexto escolar (*pp. 17-28). Havana: Pueblo y Educacion.
Lorenzo Gartia, R. and Martinez Llantada, M. (2009). *Polemics around the development of talent. In D. Castellanos Simons (Comp.) Talento: concepciones y estrategias para su desarrollo en el contexto escolar (pp. 17-28).* Havana: Pueblo y Educacion.
Llivina Lavigne, M. J. (1999). *Una Propuesta Metodologica para contribuir al desarrollo de la capacidad para resolver problemas matematicos.* Doctoral thesis, Universidad Pedagogica "Enrique Jose Varona", Havana. Cuba.
Martinez Llantada, M. (2009). *Epistemological analysis of creativity. In M. Martinez Llantada and A. Guanche Martinez (Eds.). The development of creativity. Teona y practica en la educacion (*pp. 33- 52). Havana: Pueblo y Educacion.
Masero Moreno I. C., Camacho Penalosa E. & Vazquez Cueto J. (2018). *How to assess knowledge and skills in mathematical problem solving in the economic context through rubrics?* Revista Electronica Interuniversitaria de Formation del Profesorado, 21(1), 51 - 64.
M^guez, A. (2003). *Examples, exercises, problems and questions in mathematics learning activities.* In: *Revista Educacion y Pedagog^a.* Medellin: University of Antioquia, Faculty of Education. Vol. XV, No. 35, (January-April), 2003. pp. 143 -149.
Nieto, L. J. B., Lizarazo, J. A. C & Carrasco, A. C., (2015). *Mathematics problem solving in the initial training of primary school teachers.* Spain.
OBMEP (2017). *13ª Olimpada Brasileira de Matematica das Escolas Publicas.*
OPM (2016). *Olimp^adas Portuguesas de Matematica, XXXIV edigao.*
Pacheco Urbina, V. M. (2001). *The development of outstanding talent in adolescent students. Educacion Magazine, 25(1), 123-135.*
Paiva, J., Crespo, E., & Borges, T. (2023). *Preparagao de professores para a realizagao dos concursos de Matematica na provincia do Huambo. RAC: revista angolana de ciencias. 5*(1) e050104. https://doi.org/10.54580/R0501.04

Perez Almarales, E. M. (2015). *Estrategia didactica para la preparación de concursantes en Matematica de la educacion preuniversitaria sobre la base de la gestion de conocimiento.* Havana: Universitaria.
UNDP - Angola (2002). *Os desafios pos-guerra.*
Pons, J. M. V. (2017). *Mathematical competence. Characterization of learning and evaluation activities in problem solving in compulsory education.* [Doctoral thesis, Universitat Autonoma de Barcelon, Spain].
Portugal. Ministerio da Educa^ao e Ciencia (2013). Programas e Metas Curriculares de Matematica A - Ensino Secundario.
Ramos Palacios, L. A. (2006). *Una estrategia metodologica para desarrollar olimpiadas de Matematicas en el nivel medio del sistema educativo hondureno (*Master's thesis). National Pedagogical University "Francisco Morazan".
Rodriguez, L. (2004). *Identification and evaluation of talented children.* In, M. Benavides, A. Maz Machado, E. Castro Martinez, M. R. Blanco Guijarro (Coord), La Educacion de ninos con talento en Iberoamerica (pp. 37-47). UNESCO Publishing.
Sanchez, L., Lara, L., Bravo, G., & Navales, M. (2015). *Self-preparation and reflection on teaching practice: an indispensable binomial in the pedagogical training of university teachers.* Revista de cooperacion.com, ISSN 23081953, number 7 - June.
Santos, L. M. (2014). *Mathematical Problem Solving. Fundamentos cognitivos.* Mexico. Editorial Trillas.
Sierra, Y. C. (2013). *Preparing teachers to lead the teaching-learning process using levels of assimilation.* Edumecentro, 5(2), 95-107.
Susana ,m. A & Jhonny a. A. (2014). *Talento matematico y su atención en el Ecuador.*
Vazquez, L. (2002). La preparación para los concursos de conocimiento y habilidades de F^sica en bachillerato a la luz de la educacion para la diversidad. I *Conferencia Nacional de Educacion para la Diversidad a las puestas del Siglo XXI.* Camaguey: University of Pedagogical Sciences "Jose MarU".
Villarraga, M. E., & Martinez, P. (2004). *Towards the definition of the term talent.* In, M. Benavides, A. Maz Machado, E. Castro Martinez, & M. R. Blanco Guijarro (Coord), La Educacion de ninos con talento en Iberoamerica (pp. 25-35). UNESCO Publishing.

Printed by Books on Demand GmbH, Norderstedt / Germany